McGraw-Hill My Math

Welcome to *My Math*—your very own math book!
You can write in it—in fact, you are encouraged to
write, draw, circle, explain, and color as you explore
the exciting world of mathematics. Let's get started.
Grab a pencil and finish each sentence.

My name is _____.

My favorite color is _____.

My favorite hobby or sport is _____.

My favorite TV program or video game is

_____.

My favorite class is _____.

Mc
Graw
Hill
Education

mhmymath.com

STEM McGraw-Hill is committed to providing
instructional materials in Science, Technology, Engineering,
and Mathematics (STEM) that give all students a solid
foundation, one that prepares them for college and careers
in the 21st century.

Send all inquiries to:
McGraw-Hill Education
8787 Orion Place
Columbus, OH 43240

ISBN: 978-0-07-905768-6 (**Volume 2**)
MHID: 0-07-905768-3

Printed in the United States of America.

4 5 6 7 8 9 LWI 23 22 21 20 19 18

Understanding by Design® is a registered trademark of the Association for Supervision and
Curriculum Development ("ASCD").

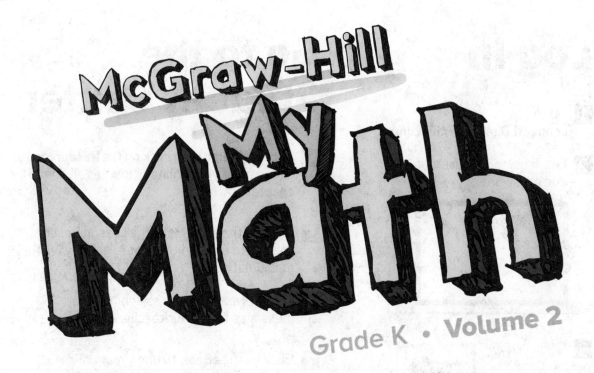

McGraw-Hill

My Math

Grade K • Volume 2

Authors:

Carter • Cuevas • Day • Malloy

Altieri • Balka • Gonsalves • Grace • Krulik • Molix-Bailey

Moseley • Mowry • Myren • Price • Reynosa • Santa Cruz

Silbey • Vielhaber

Mc
Graw
Hill
Education

GO digital ▶

▶ Log In

1 Go to **connectED.mcgraw-hill.com**.

2 Log in using your username and password.

3 Click on the Student Edition icon to open the Student Center.

Grade K

▶ Go to the Student Center

4 Click on Menu, then click on the **Resources** tab to see all of your online resources arranged by chapter and lesson.

5 Click on the **eToolkit** in the Lesson Resources section to open a library of eTools and virtual manipulatives.

6 Look here to find any assignments or messages from your teacher.

7 Click on the **eBook** to open your online Student Edition.

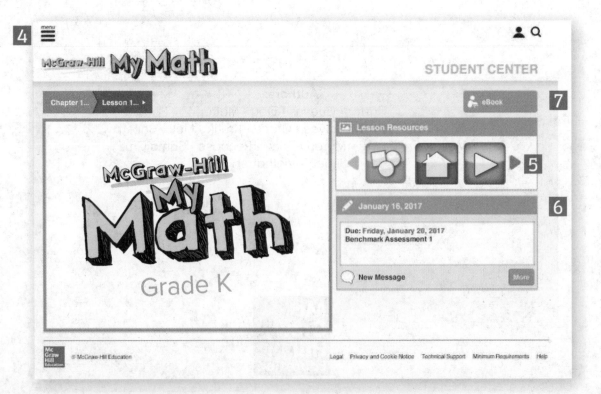

▶ Explore the eBook!

8 Click the **speaker icon** at the top of the eBook pages to hear the page read aloud to you.

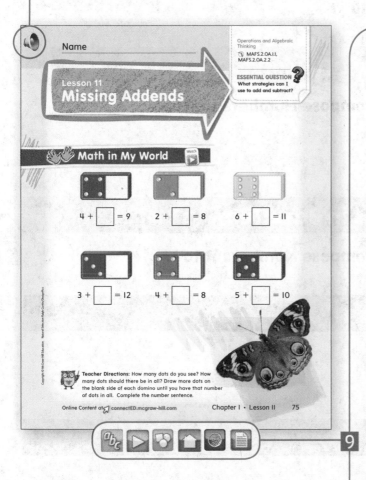

Name _____

Operations and Algebraic Thinking

MAFS.2.OA.1.1, MAFS.2.OA.2.2

ESSENTIAL QUESTION
What strategies can I use to add and subtract?

Lesson 11
Missing Addends

Math in My World

$4 + \boxed{} = 9$ $2 + \boxed{} = 8$ $6 + \boxed{} = 11$

$3 + \boxed{} = 12$ $4 + \boxed{} = 8$ $5 + \boxed{} = 10$

Teacher Directions: How many dots do you see? How many dots should there be in all? Draw more dots on the blank side of each domino until you have that number of dots in all. Complete the number sentence.

Online Content at connectED.mcgraw-hill.com   Chapter 1 · Lesson 11   75

More resources can be found by clicking the icons at the bottom of the eBook pages.

 Practice and review your Vocabulary.

 Animations and videos allow you to explore mathematical topics.

 Explore concepts with eTools and virtual manipulatives.

 eHelp helps you complete your homework.

 Explore these fun digital activities to practice what you learned in the classroom.

9

 Worksheets are PDFs for Math at Home, Problem of the Day, and Fluency Practice.

v

Contents in Brief
Organized by Domain

Processes
&Practices   Woven Throughout

Getting Started

Lessons and Homework

Wrap Up

There are *Brain Builders* problems in every lesson.

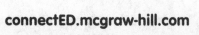

Chapter 2 Numbers to 10

Getting Started

Lessons and Homework

Wrap Up

We like healthful food!

connectED.mcgraw-hill.com

Chapter

3 Numbers Beyond 10

Getting Started

Lessons and Homework

Wrap Up

eHelp **Look for this!** Click online and you can get more help while doing your homework.

connectED.mcgraw-hill.com

Chapter 4
Compose and Decompose Numbers to 10

ESSENTIAL QUESTION
How can we show a number in other ways?

Getting Started

Lessons and Homework

Wrap Up

connectED.mcgraw-hill.com

Operations and Algebraic Thinking

ESSENTIAL QUESTION
How can I use objects to add?

Getting Started

Lessons and Homework

Wrap Up

Party time!

Look for this!
Click online and you can find tools that will help you explore concepts.

6 Subtraction

Operations and Algebraic Thinking

ESSENTIAL QUESTION
How can I use objects to subtract?

Getting Started

Lessons and Homework

Wrap Up

connectED.mcgraw-hill.com

Chapter 7 Compose and Decompose Numbers 11 to 19

Look for this!
Click online and you can find activities to help build your vocabulary.

Let it snow!

8 Measurement

ESSENTIAL QUESTION
How do I describe and compare objects by length, height, and weight?

Getting Started

Lessons and Homework

Wrap Up

 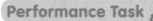

connectED.mcgraw-hill.com

Getting Started

Lessons and Homework

Wrap Up

Bright idea!

Chapter 10 Position

Getting Started

Lessons and Homework

Wrap Up

Animals in action!

connectED.mcgraw-hill.com

Copyright © McGraw-Hill Education Tetra Images/Alamy

Chapter 11 Two-Dimensional Shapes

Geometry

ESSENTIAL QUESTION
How can I compare shapes?

Getting Started

Lessons and Homework

Wrap Up

Let's learn shapes!

Geometry

ESSENTIAL QUESTION
How do I identify and compare three-dimensional shapes?

Getting Started

Lessons and Homework

Wrap Up

Shapes are fun!

connectED.mcgraw-hill.com

Chapter

6

Subtraction

Copyright © McGraw-Hill Education (l)ImageDJ/Alamy,(r)Comstock Images/Alamy

ESSENTIAL QUESTION

How can I use objects to subtract?

Let's Get Fit!

Watch ▶

Watch a video!

Name

Chapter 6 Project

Illustrate Subtraction Number Stories

1 Decide how to design the cover of your subtraction storybook.

2 Write a title on your book.

3 Use drawings and words to design your cover below.

Name ....

1

2

3

------   ------

------ and ------

Directions: 1–2. Count the objects. Write the number. **3.** Trace the number.
Count the objects. Circle the objects to show a way to take apart the
number. Write the numbers.

My Math Words

 Vocab

Review Vocabulary

 Directions: Count the children in green. Write the number. Trace the plus sign. Count the children in blue. Write the number. Trace the equals sign. Write the number that tells how many children there are in all.

Vocab Processes &Practices

are left

4 are left

minus sign (—)

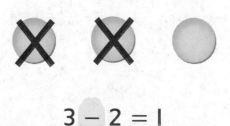

3 − 2 = 1

subtract

5 take away 3 is 2

take away

Teacher Directions:
Ideas for Use

- Have students count the letters in each word.

- Have students use the blank cards to write a word from a previous chapter that they would like to review.

minus sign

are left

take away

subtract

My Foldable

FOLDABLES® Follow the steps on the back to make your Foldable.

✂ -

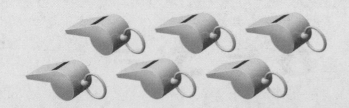

$$6 - 3 = \_\_\_$$

$$8 - 7 = \_\_\_$$

$$9 - 5 = \_\_\_$$

$$10 - 4 = \_\_\_$$

$$7 - 2 = \_\_\_$$

Lesson 1
Subtraction Stories

ESSENTIAL QUESTION
How can I use objects to subtract?

 Math in My World

 Teacher Directions: Use ⬤ to model the subtraction story. There are six pieces of food in the bowl. A dog eats three pieces. Trace the counters to model the story. Draw an X on three pieces. How many pieces are left?

1 take away ___ are left

Directions: Use counters to model each subtraction story. Trace the counters. **1.** Four boxes of cereal are on the belt. One box is taken away. Draw an X on one box. How many boxes are left? **2.** Three cans of corn are on the shelf. Two cans are taken off the shelf. Draw an X on two cans. How many cans are left?

Independent Practice

3

4

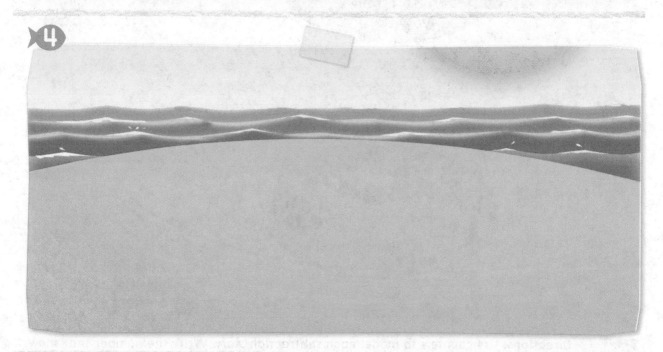

Directions: Use counters to model each subtraction story. Trace the counters. **3.** Six carrots are growing in the garden. A rabbit eats three of the carrots. Draw an X on three carrots. How many carrots are left? **4.** Four crabs are in the sand. Two of the crabs go in the water. Draw an X on two crabs. How many crabs are left?

5

_____

_____

_____

6

_____

_____

_____

7

_____

_____

_____

 Directions: Use counters to model each subtraction story. Write the number that shows how many are left. **5.** There are three pieces of food. Two pieces are eaten. How many pieces are left? **6.** There are seven pieces of food. Two pieces are eaten. How many pieces are left? **7.** There are ten pieces of food. One piece is eaten. How many pieces are left? Two more pieces of food are eaten. Tell a friend how many pieces of food are left now. Tell how many total pieces of food were subtracted from ten.

My Homework

Homework Helper

Need help? ↗ connectED.mcgraw-hill.com

1

2

 Directions: Use pennies to model each subtraction story. Trace the pennies. **1.** Three people are in the park. Two leave. Draw an X on two people. How many people are left? **2.** There are six balls. Four roll away. Draw an X on four balls. How many balls are left?

Vocabulary Check

4 take away   are left

_____

 Directions: 3. Use pennies to model the subtraction story. Trace the pennies. Four people are on the trail. Three people walk away. Draw an X on three people. How many people are left? **4.** I have six balloons. Three of the balloons blow away. How many balloons are left? Write the number.

Math at Home   Tell your child a subtraction story. Have your child use pennies to model the story. Ask them to tell you how many are left.

Name _____

ESSENTIAL QUESTION
How can I use objects to subtract?

👐 **Math in My World**  Tools | Watch

_____ **are left**

🦉 **Teacher Directions:** Use ⬤ to model the subtraction story. Six astronauts landed on the moon. Two astronauts leave. Trace the counters to model your story. Draw an X on two astronauts. Write the number that tells how many astronauts are left.

1

2 are left

2

_____ are left

3

_____ are left

Directions: Use counters to model subtraction. **1.** Three penguins are on an iceberg. One penguin got tired and went away. How many penguins are left? **2.** Six snowflakes fell from the sky. Three melted. How many snowflakes are left? **3.** Three snowballs are stacked to make a snowperson. Two snowballs fell off. How many are left?

Name _____

Independent Practice

4 _____ are left

5 _____ are left

6 _____ are left

Directions: Use counters to model subtraction. **4.** Five penguins are swimming together. Two swim away. How many penguins are left? **5.** Seven penguins are in an igloo. Three penguins go out to play. How many penguins are left? **6.** A penguin has six mittens. She loses one. How many mittens are left?

Directions: 7. Listen to the subtraction story. There were six lions sleeping. Four lions woke up. Draw an X on the lions that woke up. Tell how many lions are still sleeping. Write the number. Another lion woke up. Tell a friend how many lions are still sleeping now. Tell your friend how many lions have woken up in all.

My Homework

Homework Helper eHelp

Need help? connectED.mcgraw-hill.com

1 **2** are left

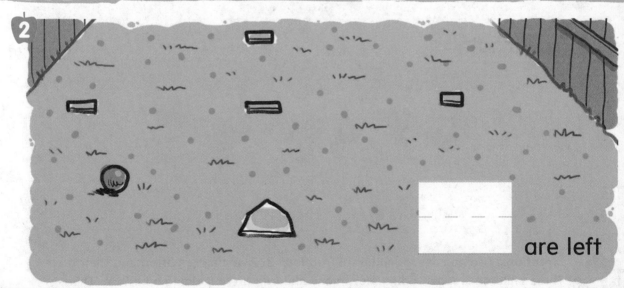

2 ___ are left

Directions: Use pennies to model subtraction. **1.** Three children are on the balance beam. One child went to another activity. How many children are left? **2.** Eight children are playing kickball. Four children go home. How many children are left?

☐ ☐ **are left**

Vocabulary Check

🐟 **4** subtract

_____

- - - - -

_____

 Directions: Use pennies to model subtraction. **3.** Five children are riding their bikes. Three children stopped riding. How many children are left riding bikes? **4.** There are four skateboards. One skateboard rolls away. Draw an X on one skateboard. Tell how many are left. Write the number.

Math at Home Have your child tell you a short subtraction story and use buttons or dry cereal to model the story.

Name

....

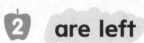

Check My Progress

Vocabulary Check

1 take away

2 are left

_____

\- \- \- \- \-

_____ are left

Concept Check

3

_____

\- \- \- \- \-

_____

 Directions: Use counters to model each subtraction story. Trace the counters to show your work. **1.** Five rabbits are in the grass. Three rabbits hop away. How many rabbits are left? **2.** Four butterflies are on a branch. One butterfly flies away. How many butterflies are left? Write the number. **3.** Eight sailboats are on the water. Five sailboats sail away. How many sailboats are left? Write the number.

_____

_ _ _ _ _

_____ are left

_____

_ _ _ _ _

_____ are left

Directions: Use counters to model each subtraction story. Trace the counters to show your work. **4.** There are five books on the shelf. Three books are taken. Draw an X on three books. How many books are left? **5.** There are four frogs on a log. Two frogs hop off. How many frogs are left? Write the number. **6.** There are seven toys in the wagon. Four toys are taken. How many toys are left? Write the number.

Lesson 3
Use the − Symbol

ESSENTIAL QUESTION
How can I use objects to subtract?

 Math in My World Tools Watch

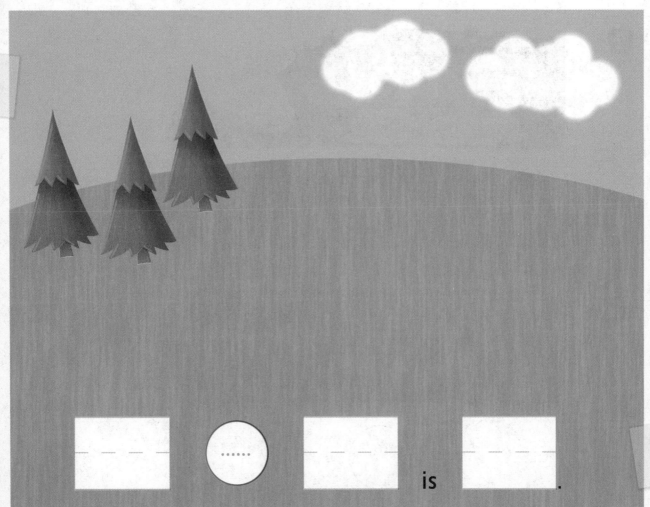

is   .

 Teacher Directions: Use ⬤ to model the subtraction story. Six children are playing. Trace the counters. Write the number. Trace the minus sign. Four children have to go home. Write the number that tells how many children go home. Draw an X on the counters that were taken away. Write the number that tells how many are left.

1

minus sign (−)

5 (−) 2 is 3.

2

(.....) is ____.

3

(.....) is ____.

Directions: 1–3. Count and write the number of animals. Trace the minus sign. Draw an X on the animals that are going away. Write the number that tells how many animals are going away. Write the number that tells how many are left.

Name

4

_____ _____

(……)

_____ **is** _____ .

5

_____ _____

(……)

_____ **is** _____ .

6

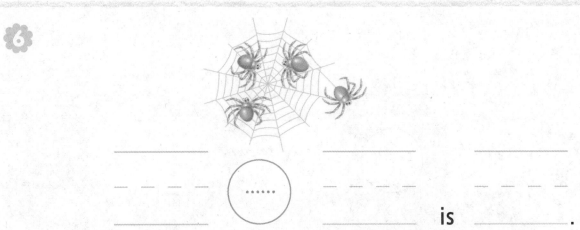

_____ _____

(……)

_____ **is** _____ .

Directions: 4–6. Count and write the number of animals. Trace the minus sign. Draw an X on the animals that are going away. Write the number that tells how many animals are going away. Write the number that tells how many are left.

 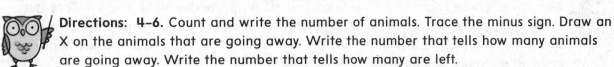

7

_____ ⊙····· _____ is _____.

 Directions: 7. Use counters to create and model a subtraction story. Trace the counters. Write the number. Trace the minus sign. Draw an X on the counters that are taken away. Write the number. Tell how many are left. Write the number. Explain to a friend how you know when to use a minus sign.

My Homework

Homework Helper eHelp

Need help? connectED.mcgraw-hill.com

1

$$6 \ominus 2 \quad \text{is} \quad 4 \, .$$

2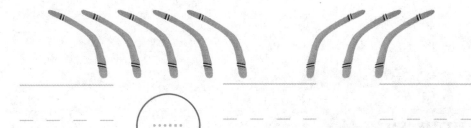

_____ ⬭ _____ is _____ .

3

_____ ⬭ _____ is _____ .

 Directions: 1–3. Count and write the number of objects. Trace the minus sign. Draw an X on the objects that are going away. Write the number that tells how many are going away. Write the number that tells how many are left.

Chapter 6 • Lesson 3   401

4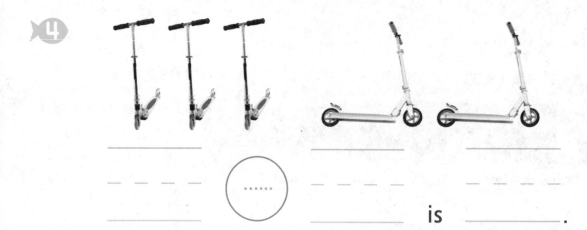

_____

_____   (......)   _____   is _____ .

5

_____

_____   (......)   _____   is _____ .

Vocabulary Check

6 **minus sign (–)**

7   (......)   3   is _____ .

Directions: 4–5. Count and write the number of objects. Trace the minus sign. Draw an X on the objects that are going away. Write the number that tells how many are going away. Write the number that tells how many are left. **6.** There are seven tennis rackets. Three are being used. Draw an X on each racket that is being used. Trace the minus sign. Tell how many are left. Write the number.

Math at Home   Draw 9 squares on paper. Have your child put an X on 6 of the squares. Have your child write the number that tells how many are left.

Name

....

Lesson 4
Use the = Symbol

ESSENTIAL QUESTION
How can I use objects to subtract?

 Math in My World   Tools   Watch

HOME SWEET HOME

[   ] — [   ] (......) [   ]

 Teacher Directions: Use ⬤ to model the subtraction story. Five stuffed animals are on the table. Write the number. Three stuffed animals are moved to the rug. Write the number. Trace the equals sign. Write how many stuffed animals are left on the table.

Processes & Practices

1

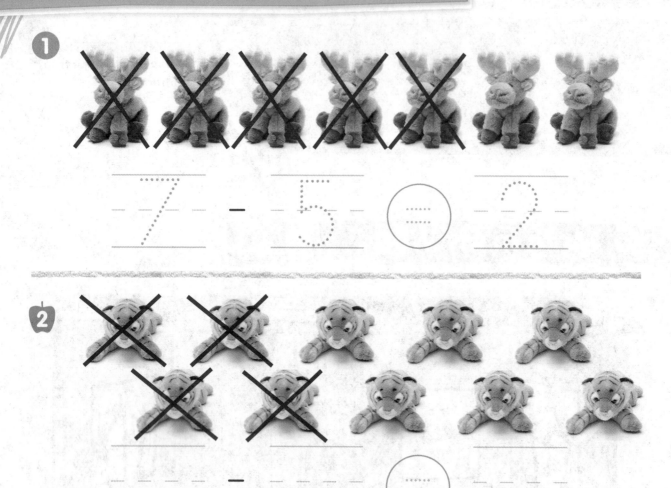

7 − 5 ⊜ 2

2

_____ − _____ ⊜ _____

3

_____ − _____ ⊜ _____

Directions: 1–3. Count and write the number of toys. Write the number that tells how many toys are taken away. Trace the equals sign. Write the number that tells how many toys are left.

Independent Practice

4 _ _ _ _ _ − _ _ _ _ _ ⊙ _ _ _ _ _

5 _ _ _ _ _ − _ _ _ _ _ ⊙ _ _ _ _ _

6 _ _ _ _ _ − _ _ _ _ _ ⊙ _ _ _ _ _

Directions: 4–6. Count and write the number of toys. Write the number that tells how many toys are taken away. Trace the equals sign. Write the number that tells how many toys are left.

 Directions: 7. Count the dolls in the window. Write the number of dolls. Take some dolls away. Draw an X on the dolls that were taken away. Write the number. Trace the equals sign. Tell how many are left. Write the number. Tell a friend when you used the equals sign before. Explain why you think it works for subtraction, too.

My Homework

Homework Helper

Need help? connectED.mcgraw-hill.com

1

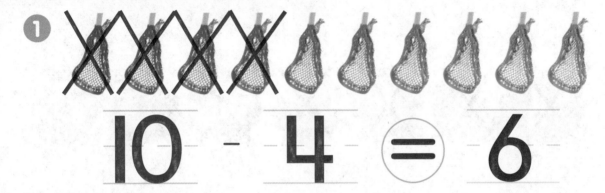

$$10 - 4 = 6$$

2

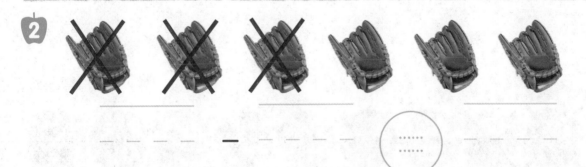

___ − ___ ⬭

3

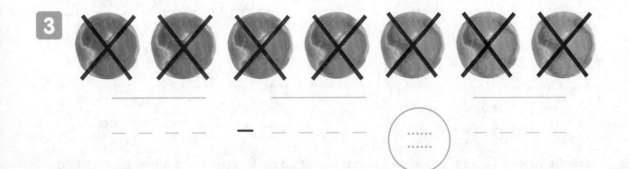

___ − ___ ⬭

Directions: 1–3. Count and write the number of objects. Write the number that tells how many are taken away. Trace the equals sign. Write how many objects are left.

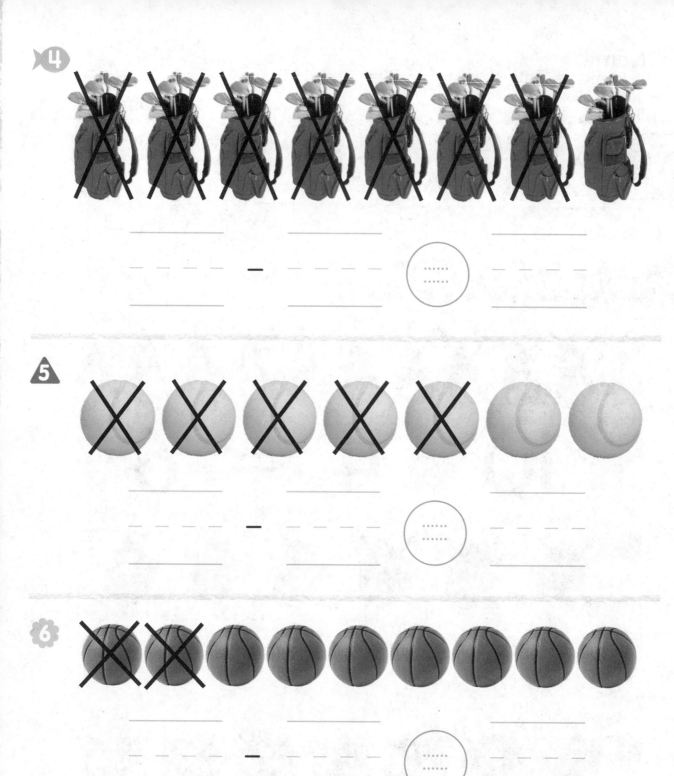

4

_____ **−** _____ **⊙** _____

5

_____ **−** _____ **⊙** _____

6

_____ **−** _____ **⊙** _____

Directions: 4–6. Count and write the number of objects. Write the number that tells how many are taken away. Trace the equals sign. Write how many objects are left.

Math at Home Show your child 7 objects. Have your child write the number of objects. Write the minus sign. Have your child take 4 of the objects away. Write the number sentence. Have your child write the number that tells how many are left.

Name _____

Math in My World

Tools | Watch

Teacher Directions: Use ⬤ to model the subtraction story. Write how many bales of hay. Trace the minus sign. A farmer takes away four bales. Draw an X on those bales. Write the number. Trace the equals sign. Write how many are left.

Processes &Practices

1

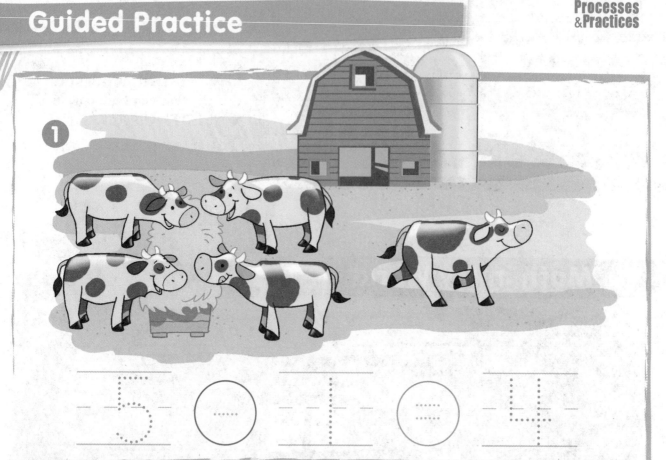

5 ⬚ − ⬚ 4

2

Directions: 1–2. Count and write the number of animals. Trace the minus sign. Draw an X on the animals going away. Write the number. Trace the equals sign. Write the number that tells how many animals are left.

Name

_____

Independent Practice

3

_____ (......) _____ (......) _____

4

_____ (......) _____ (......) _____

5

_____ (......) _____ (......) _____

Directions: 3–5. Count and write the number of animals. Trace the minus sign. Draw an X on the animals going away. Write the number. Trace the equals sign. Write the number that tells how many animals are left.

Directions: 6. Use counters to create and model a subtraction story. Trace the counters used. Write the number of counters used. Trace the minus sign. Draw an X on the counters that are taken away. Write the number. Trace the equals sign. Write the number that tells how many are left. Explain to a friend how taking away a counter and drawing an X on the counter are similar. Tell your friend which you prefer and why.

Name _____

My Homework

Homework Helper

Need help? connectED.mcgraw-hill.com

1

$$7 - 4 = 3$$

2

◯ ...... ◯ ......

 Directions: 1–2. Count and write the number of animals. Trace the minus sign. Draw an X on the animals going away. Write the number. Trace the equals sign. Write the number that tells how many animals are left.

3

_____ ____ ⬭ ____ ____ ⬭ ____ _____

4

_____ ____ ⬭ ____ ____ ⬭ ____ _____

Directions: 3–4. Count and write the number of animals. Trace the minus sign. Draw an X on the animals going away. Write the number. Trace the equals sign. Write the number that tells how many animals are left.

Math at Home   Show 10 crayons. Have your child take 6 away. Have your child write a number sentence to show how many are left.

414   **Chapter 6 · Lesson 5**

Name
....

Lesson 6
Problem Solving
STRATEGY: Write a Number Sentence

How many are left?

Write a Number Sentence

$$10 \; \bigcirc \; 3 \; \bigcirc \; 7$$

 Teacher Directions: Use to model the subtraction story. There are 10 suitcases on the truck. Three suitcases are moved to the ground. Trace the number sentence to show how many suitcases are left on the truck.

How many are left?

Write a Number Sentence

 Directions: Use counters to model the subtraction story. My aunt has eight quilt squares to sew on the quilt. She sews six squares. Write a number sentence to show how many squares are left to sew on the quilt.

Name

How many are left?

Write a Number Sentence

___ ___ ___ ___ ___ (......) ___ ___ ___ ___ (......)

 Directions: Use counters to model the subtraction story. There are nine toys. Four are put on the shelf. Write a number sentence to show how many toys are left to put on the shelf.

How many are left?

Write a Number Sentence

 Directions: Use counters to model the subtraction story. There are five playground balls outside. The children are using two of the balls. Write a number sentence to show how many playground balls are left.

My Homework

How many are left?

Write a Number Sentence

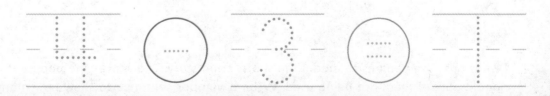

4 ◯ 3 ◯ 1

Directions: Use pennies to model the subtraction story. There are four hula hoops. Three hula hoops are being used. Trace the number sentence to show how many hula hoops are left.

How many are left?

Write a Number Sentence

_____ _____ _____

Directions: Use pennies to model the subtraction story. There are nine pool noodles. Six of them are being used. Write a number sentence to show how many are left.

Math at Home Take advantage of problem-solving opportunities during daily routines such as riding in the car, bedtime, doing laundry, putting away groceries, and so on.

Name ....

ESSENTIAL QUESTION
How can I use objects to subtract?

 Math in My World [Tools]

Teacher Directions: Use ⬤ to model taking away from 10. There are 10 players. Trace the number and minus sign. Four players leave. Write the number. Trace the equals sign. How many players are left on the court? Write the number.

1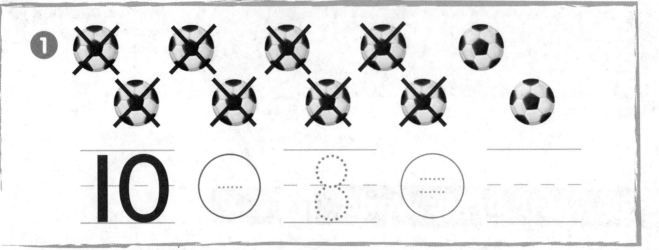

$$10 \quad \bigcirc \quad 8 \quad \bigcirc$$

2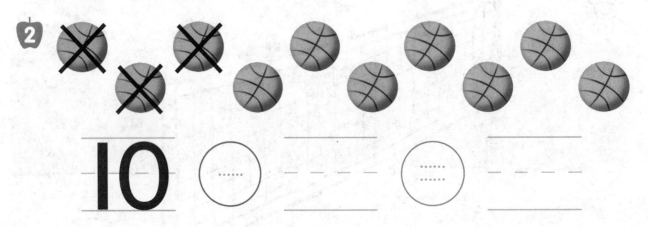

$$10 \quad \bigcirc \quad \quad \bigcirc$$

3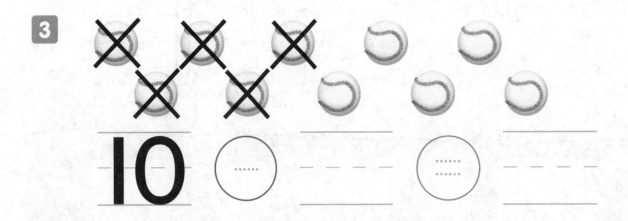

$$10 \quad \bigcirc \quad \quad \bigcirc$$

Directions: Use counters to model taking away from 10. **1.** Count the balls. Trace the minus sign. Trace the number that is taken away. Trace the equals sign. Write how many balls are left. **2-3.** Count the balls. Trace the minus sign. Write how many are taken away. Trace the equals sign. Write how many balls are left.

Name

Independent Practice

4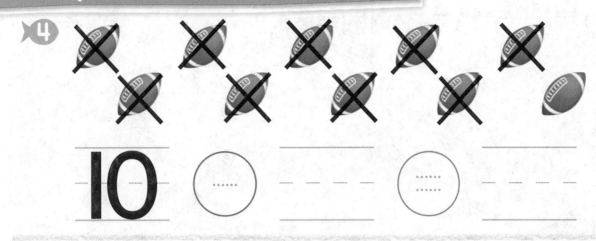

10 — (⋯⋯) _____ (⋯⋯) _____

5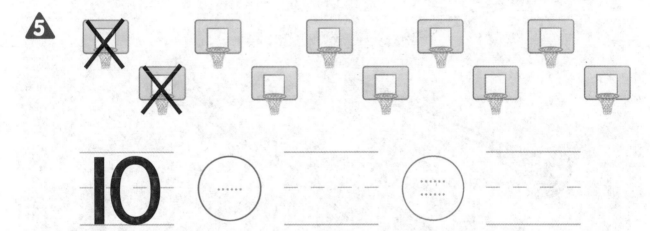

10 — (⋯⋯) _____ (⋯⋯) _____

6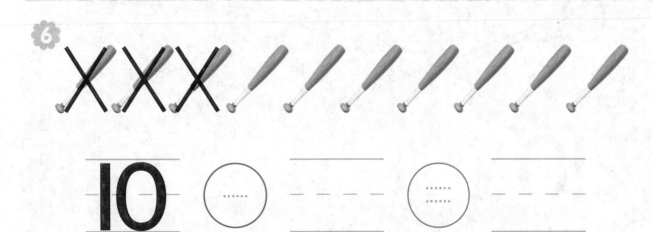

10 — (⋯⋯) _____ (⋯⋯) _____

 Directions: 4–6. Use counters to model taking away from 10. Count the objects. Trace the minus sign. Write how many are taken away. Trace the equals sign. Write how many are left.

Copyright © McGraw-Hill Education.

Name _____

Independent Practice

4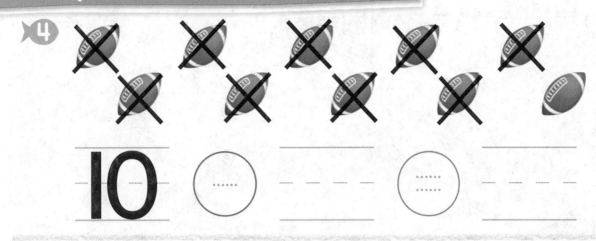

10 — (····) _____ (····) _____

5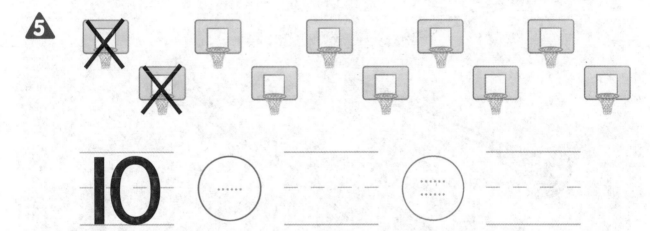

10 — (····) _____ (····) _____

6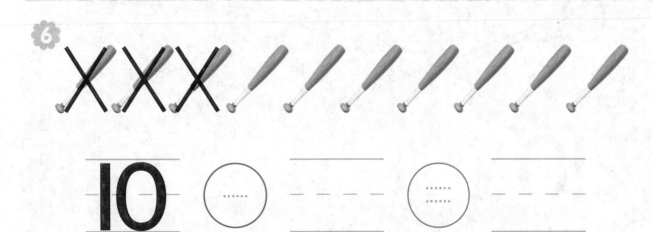

10 — (····) _____ (····) _____

Copyright © McGraw-Hill Education.

 Directions: 4–6. Use counters to model taking away from 10. Count the objects. Trace the minus sign. Write how many are taken away. Trace the equals sign. Write how many are left.

Online Content at connectED.mcgraw-hill.com

Chapter 6 • Lesson 7

423

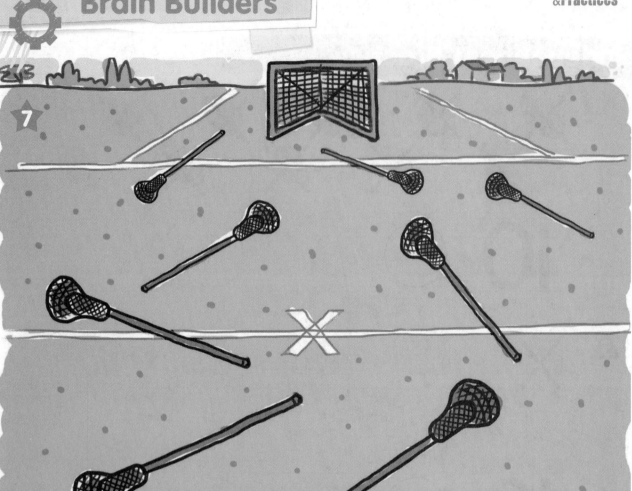

Directions: 7. Use counters to model taking away from 10. There are 10 lacrosse sticks. Trace the minus sign. Only six players showed up to play. Draw an X on six sticks. Write the number. Trace the equals sign. Write how many sticks are left. Tell a friend how writing a number sentence can help you solve the problem.

My Homework

Homework Helper

Need help? connectED.mcgraw-hill.com

1

10 — 9 = 1

2

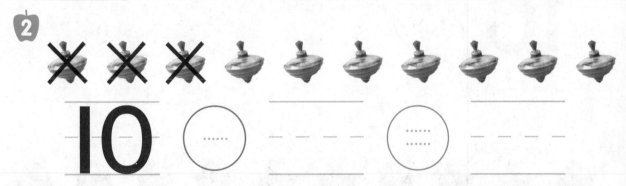

10 ⊖ ⸱⸱⸱⸱ _____ ⸱⸱⸱⸱

3

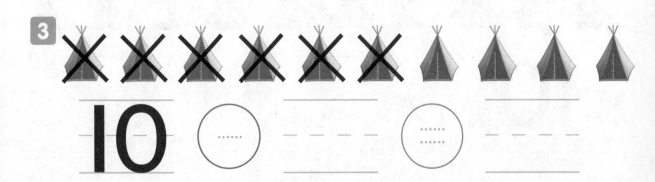

10 ⊖ ⸱⸱⸱⸱ _____ ⸱⸱⸱⸱

Directions: 1–3. Use pennies to model taking away from 10. Trace the minus sign. Write how many are taken away. Trace the equals sign. Write how many are left.

4

10 ◯ - - - - ◯ - - - -

5

10 ◯ - - - - ◯ - - - -

6

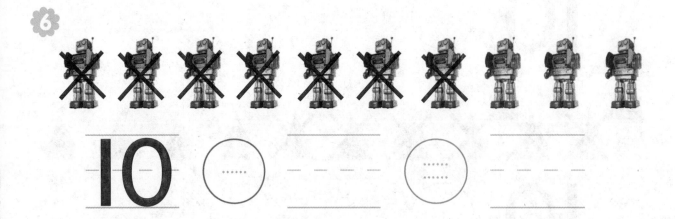

10 ◯ - - - - ◯ - - - -

Directions: 4–6. Use pennies to model taking away from 10. Trace the minus sign. Write how many are taken away. Trace the equals sign. Write how many are left.

Math at Home Give your child ten objects. Have your child practice taking away from 10.

Name
....

Fluency Practice

1

_____ _____ (......) _____ - _____ (......) _____ = _____

2

_____ _____ (......) _____ - _____ (......) _____ = _____

3

_____ _____ (......) _____ - _____ (......) _____ = _____

Directions: 1–3. Count the objects in each group. Write the number. Trace the minus sign. Count the objects that have been taken away. Write the number. Trace the equals sign. Write how many there are left.

Fluency Practice

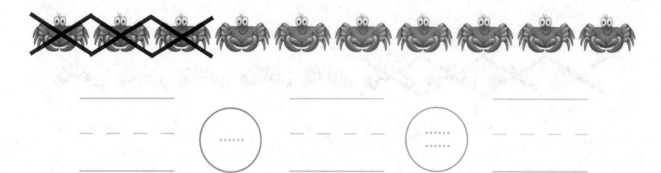

_____ (.....) _____ (.....) _____

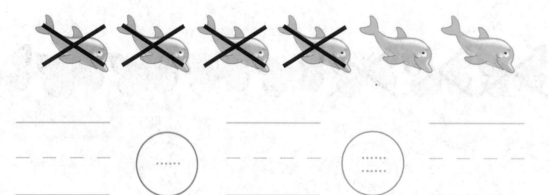

_____ (.....) _____ (.....) _____

_____ (.....) _____ (.....) _____

Directions: 4–6. Count the objects in each group. Write the number. Trace the minus sign. Count the objects that have been taken away. Write the number. Trace the equals sign. Write how many there are left.

Name _____

My Review

Vocabulary Check

1 minus sign

5 take away 2

2 equals sign

—

3 subtract

=

4 are left

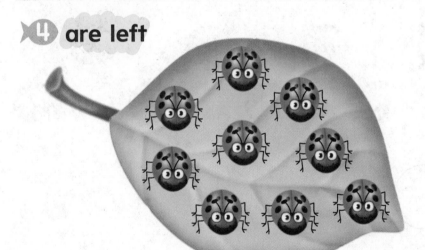

_____

_ _ _ _ _ _ _ _

_____ are left

 Directions: 1–3. Draw lines to match each word with the correct symbol or picture.
4. Nine ladybugs are on the leaf. Four ladybugs flew away. Draw an X on each
ladybug that flew away. How many ladybugs are left? Write the number.

Concept Check

5

_____ are left

6

_____ ◯.... ◯...... _____

7

IO ◯.... ◯......

 Directions: 5. Use counters to model the subtraction story. Trace the counters. Five spiders are on a web. Three spiders crawl away. Write how many are left. **6.** Count and write the number of stickers. Trace the minus sign. Write the number of stickers that are taken away. Trace the equals sign. Write how many are left. **7.** Count the objects. Trace the minus sign. Write how many are taken away. Trace the equals sign. Write how many are left.

Brain Builders

8

10   ○   _____   ○   _____

Directions: 8. Use counters to model taking away from 10. There are 10 balls in the container. Trace the counters to show ten balls. Seven were used during gym class. Draw an X on seven counters. Write the number. Trace the minus sign. Trace the equals sign. Write how many are left.

 Directions: Count the lions. Say the number. Trace the minus sign. Count the lions that are sleeping. Draw an X on each sleeping lion. Write the number. Trace the equals sign. Write how many lions are awake.

Performance Task

Field Day at School

The school is having a Field Day. Students will play games and have lots of fun outdoors.

Show all of your work to receive full credit.

Part A

There are five bikes in a race. Two bikes get flat tires. Draw an X on two bikes. How many bikes are left in the race? Write the number sentence.

_____ _____ _____

_____ — _____ = _____

Part B

There are seven balloons. One balloon bursts. Circle the number that shows how many balloons are left.

4   5   6

Part C

There are eight kites. Some kites fly away. Write a number sentence to find how many kites are left.

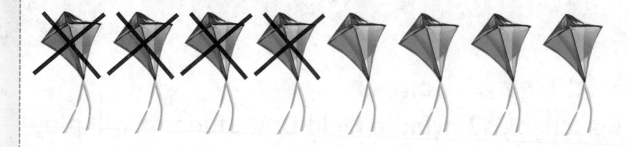

_____ − _____ = _____

Part D

Trophies were given for first place. Circle the number sentence that shows how many trophies were given away.

10 − 7 = 3   9 − 7 = 2   10 − 8 = 2

Chapter 7

Compose and Decompose Numbers 11 to 19

ESSENTIAL QUESTION

How do we show numbers 11 to 19 in another way?

Look at Our Changing Seasons!

Watch a video!

Chapter 7 Project

Collections Poster

1 Write the number your group chose to take apart.

2 Draw the total number in the ten-frames.

3 Circle a group of objects that shows 10. Write the number.

4 Circle the group that shows some more. Write the number.

5 Choose a new number to take apart. Repeat.

_____ and _____ more

_____ and _____ more

Name _____

1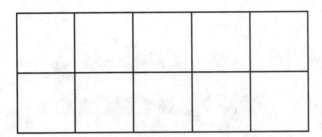

_ _ _ _ _ _

2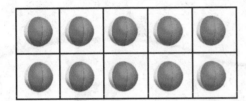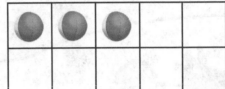

_ _ _ _ _ _

3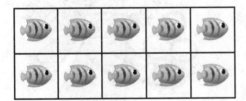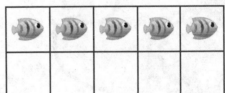

_ _ _ _ _ _

4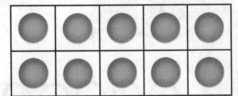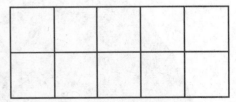

 Directions: I. Draw objects in the ten-frame to show 10. Write the number.
2–3. Count the objects. Say how many. Write the number. **4.** Say the
number. Trace the number. Draw more counters to show the number.

My Math Words

Review Vocabulary

sixteen

thirteen

Directions: Count the snowflakes. Say how many. Trace the number. Circle the group of 10 snowflakes. Count the snowballs. Say how many. Trace the number. Color the group of 10 snowballs.

My Vocabulary Cards

Vocab abc

eighteen 18

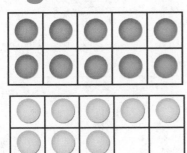

eleven 11

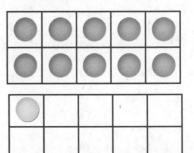

fifteen 15

fourteen 14

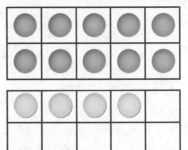

nineteen 19

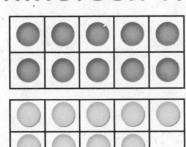

seventeen 17

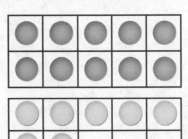

Teacher Directions:
Ideas for Use

- Instruct students to choose a card and show how many by drawing that many objects. Have a classmate count the objects and tell which card was chosen.

- Have students name the letters in each word.
- Guide students to find the number words that end with the same four letters. Have them say each number.

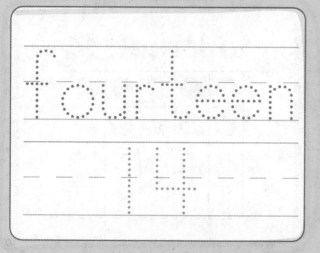

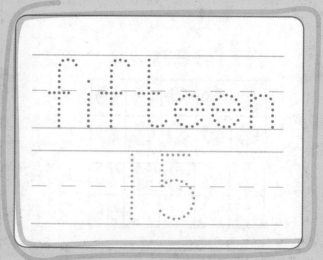

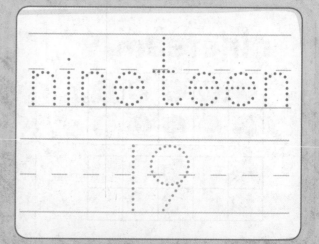

My Vocabulary Cards

Vocab abc

sixteen 16

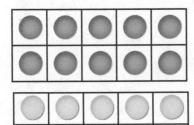

ten 10

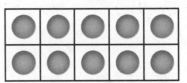

thirteen 13

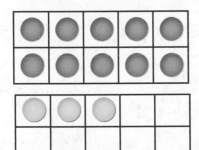

twelve 12

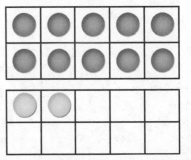

Teacher Directions:
More Ideas for Use

- Have students draw a mark on each card every time they read the word in this chapter or use it in their writing. Challenge them to make five marks for each card.

- For each blank card, guide students to design a card that shows a number less than 10. Have them pair each new card with the number 10 card. Say the number.

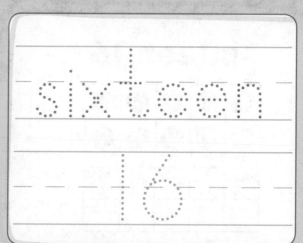

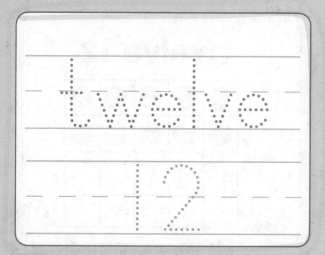

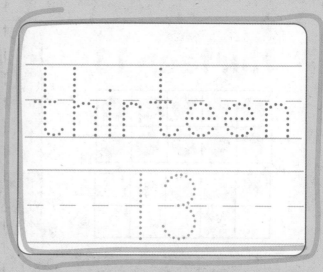

My Foldable

FOLDABLES Follow the steps on the back to make your Foldable.

✂

Take Apart Numbers

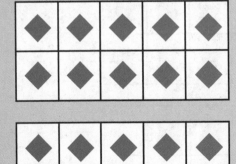

_____ and _____ more

_____ and _____ more

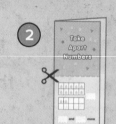
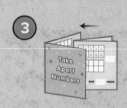

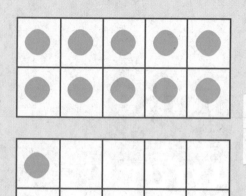

and more

and more

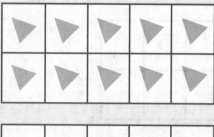

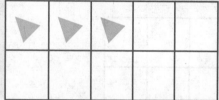
and more

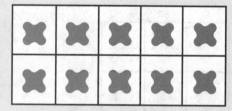
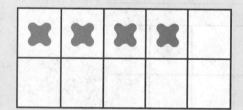
and more

Name _____

Make Numbers 11 to 15

ESSENTIAL QUESTION
How do we show numbers 11 to 19 in another way?

Math in My World

Tools Watch

Teacher Directions: Use ⬤ to show 10. Write the number. Use ⬤ to show four more. Write the number. Use red to draw counters to show 10. Use yellow to draw counters to show four more. What number does 10 and four more make? Write the number.

Processes &Practices

1

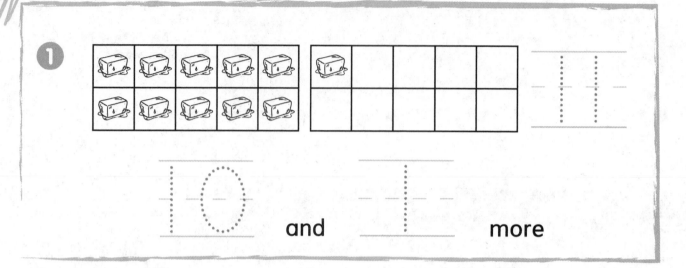

10

and

1

more

11

2

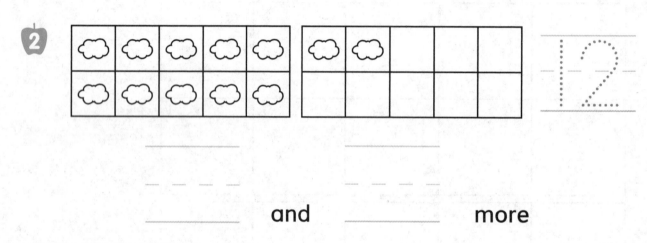

12

and

more

3

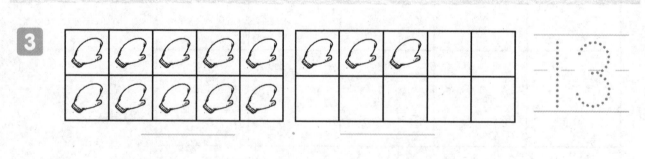

13

and

more

Directions: 1–3. Count 10. Use Work Mat 4 and red counters to show 10. Color the objects red to show 10. Write the number. Count how many more. Use yellow counters to show how many more. Color the other objects yellow to show how many more. Write the number. Trace the number made.

Name

Independent Practice

  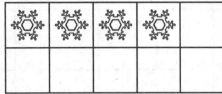 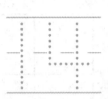

_____ and _____ more

 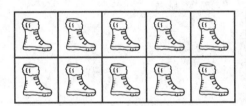 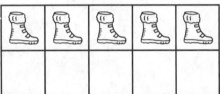

_____ and _____ more

 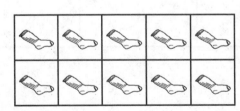 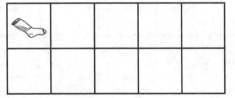

_____ and _____ more

Directions: 4–6. Count 10. Use Work Mat 4 and red counters to show 10. Color the objects red to show 10. Write the number. Count how many more. Use yellow counters to show how many more. Color the other objects yellow to show how many more. Write the number. Trace the number made.

Online Content at connectED.mcgraw-hill.com

Chapter 7 · Lesson 1   445

Copyright © McGraw-Hill Education

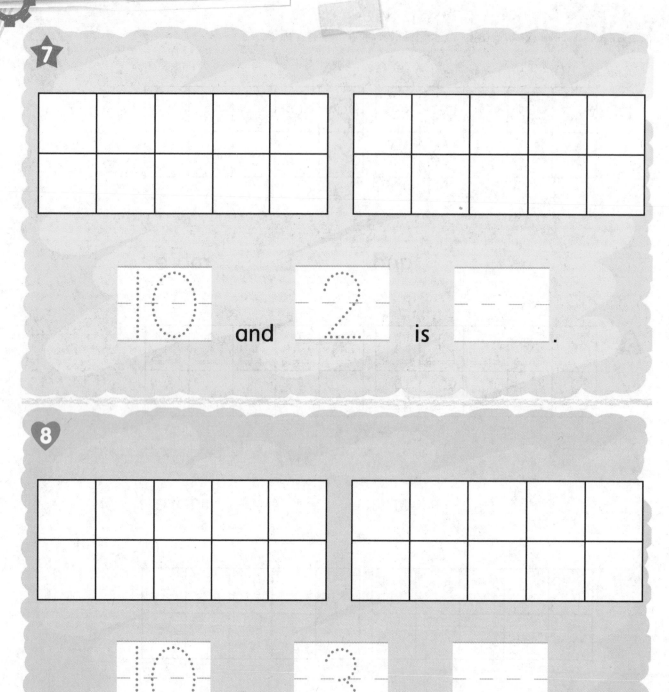

7

10   and   2   is   ___ .

8

10   and   3   is   ___ .

Directions: 7–8. Trace 10. Draw objects to show 10 in the ten-frame. Color the objects red. Trace the other number. Draw objects to show that many in the other ten-frame. Color the objects yellow. Write the number made. Tell a friend what number would be made if you would draw two more objects in the second ten-frame in Exercise 8.

Name _____

My Homework

Homework Helper

Need help? connectED.mcgraw-hill.com

1 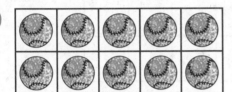   13

10 and 3 more

2 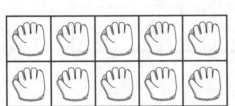   15

_____ and _____ more

Directions: 1–2. Count 10. Color the objects red to show 10. Write the number.
Count how many more. Color the other objects yellow to show how many more.
Write the number. Trace the number made.

Chapter 7 · Lesson 1   447

3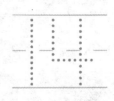

_____

_____ and _____ more

4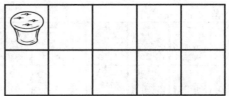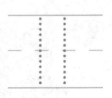

_____

_____ and _____ more

5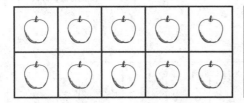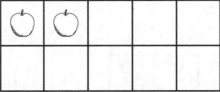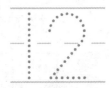

_____

_____ and _____ more

Directions: 3–5. Count 10. Color the objects red to show 10. Write the number. Count how many more. Color the other objects yellow to show how many more. Write the number. Trace the number made.

Math at Home Give your child 10 pieces of dry cereal. Give your child 3 more. Have your child make one group of 10 and 3 more. Have your child tell you how many 10 and 3 more make.

Name

Take Apart Numbers 11 to 15

ESSENTIAL QUESTION
How do we show numbers 11 to 19 in another way?

Math in My World

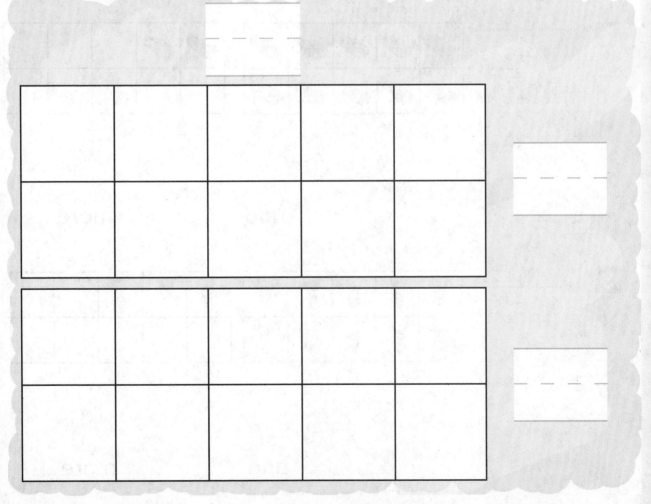

Teacher Directions: Say a number between 11 and 15. Write the number above the ten-frames. Use ⬤ to show the number. Use one ten-frame to show 10. Write the number 10. Use the other ten-frame to show the other counters. Write the number. Draw the counters in the ten-frames to show a group of 10 and some more.

Guided Practice

1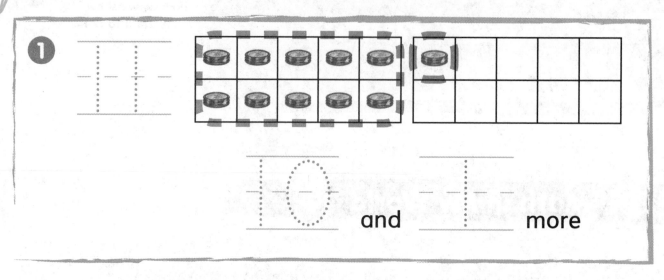

10 and 1 more

2

_____ and _____ more

3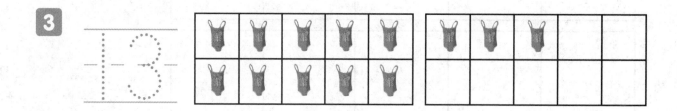

_____ and _____ more

 Directions: 1–3. Say the number. Trace it. Use Work Mat 4 and red counters to show the number. Take apart the number by showing 10 counters and some more. Write the numbers. Circle the group of objects that shows 10. Circle the group that shows some more.

Independent Practice

 4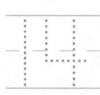

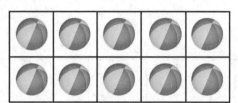 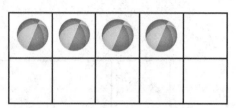

_____ and _____ more

5

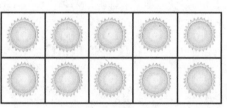 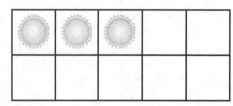

_____

_____ and _____ more

6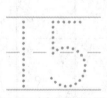

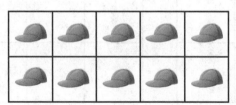 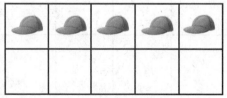

_____

_____ and _____ more

 Directions: 4–6. Say the number. Trace it. Use Work Mat 4 and red counters to show the number. Take apart the number by showing 10 counters and some more. Write the numbers. Circle the group of objects that shows 10. Circle the group that shows some more.

7

14

_____ _____   _____

_____ is _____ and _____ .

8

13

_____ _____   _____

_____ is _____ and _____ .

Directions: 7–8. Say the number. Trace it. Draw objects in the ten-frames to show the number. Circle the group of objects that shows 10. Circle the group that shows some more. Write the number sentence. Explain to a friend how to take apart a number.

Name _____

My Homework

Homework Helper eHelp

Need help? connectED.mcgraw-hill.com

1

12

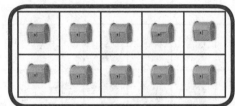

10 and 2 more

2

15

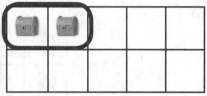

_____ and _____ more

Directions: 1–2. Say the number. Trace it. Circle the group of objects that shows 10. Write the number. Circle the group that shows some more. Write the number.

_____

_____ and _____ more

 4

_____

_____ and _____ more

 5

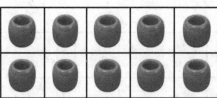

_____

_____ and _____ more

Directions: 3–5. Say the number. Trace it. Circle the group of objects that shows 10.
Write the number. Circle the group that shows some more. Write the number.

Math at Home Show your child 11 pennies. Have your child use the pennies to show
a group of 10 and some more. Repeat using 12, 13, 14, and 15 pennies.

Name _____

Lesson 3
Problem Solving
STRATEGY: Make a Table

ESSENTIAL QUESTION
How do we show numbers 11 to 19 in another way?

What numbers make 15?

Make a Table

Teacher Directions: Count how many bears. Say the number. Count the red bears. Trace the number in the table. Count the yellow bears. Trace the number in the table.

Copyright © McGraw-Hill Education

What numbers make 14?

Make a Table

Directions: Count how many color tiles. Say the number. Count the blue tiles. Write the number in the table. Count the green tiles. Write the number in the table.

What numbers make 12?

12

Make a Table

Directions: Count how many game pieces. Say the number. Count the purple pieces. Write the number in the table. Count the orange pieces. Write the number in the table. Explain to a friend how using a table is like using ten-frames.

Make a Table

 Directions: Count how many shapes. Say the number. Count the red shapes. Write the number in the table. Count the blue shapes. Write the number in the table.

My Homework

What numbers make 12?

Make a Table

 Directions: Count how many bears. Say the number. Count the orange bears. Trace the number in the table. Count the purple bears. Trace the number in the table.

What numbers make 15?

Make a Table

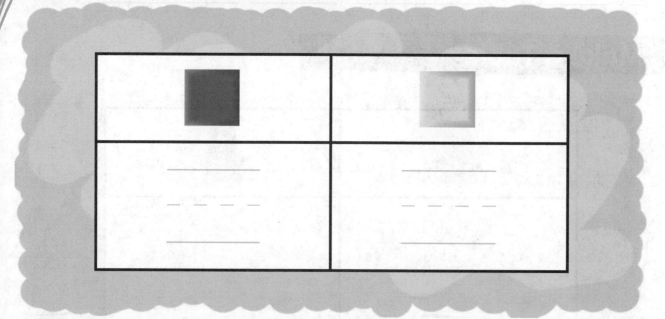

 Directions: Count how many color tiles. Say the number. Count the green tiles. Write the number in the table. Count the yellow tiles. Write the number in the table.

Math at Home   Take advantage of problem-solving opportunities during daily routines such as making breakfast. Use dry cereal pieces to show ways to make a number.

Name _____

Check My Progress

Vocabulary Check

1 **eleven 11**

2 **fifteen 15**

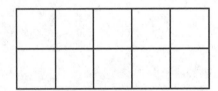

Concept Check

3

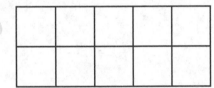

_____

- - - - - - - - - - -

_____ and _____ more

 Directions: 1. Count the objects. Draw more to show 11. **2.** Color the boxes in the ten-frames to show 15. **3.** Count 10. Color the objects red to show 10. Write the number. Count how many more. Color the objects yellow to show how many more. Write the number. Trace the number made.

4 ___ 13 ___

_____ _____

_____ and _____ more

5 ___ 12 ___

_____ _____

_____ and _____ more

6 ___ 15 ___

_____ _____

_____ and _____ more

Directions: 4–6. Say the number. Trace it. Circle the group of objects that shows 10.
Write the number. Circle the group that shows some more. Write the number.

Lesson 4
Make Numbers 16 to 19

ESSENTIAL QUESTION
How do we show numbers 11 to 19 in another way?

 Math in My World Tools Watch

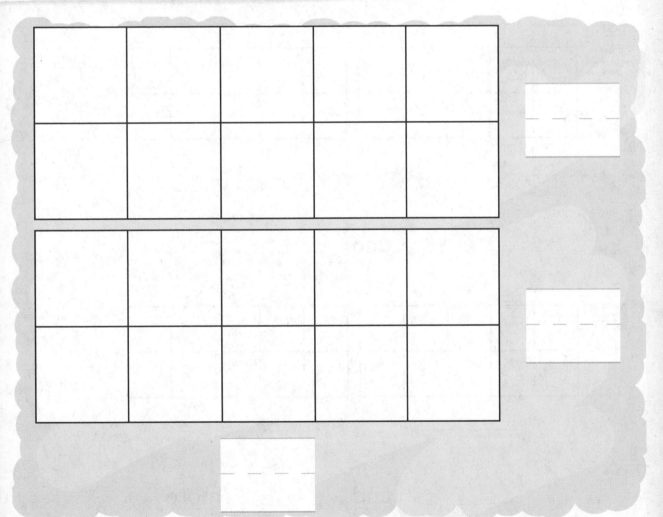

 Teacher Directions: Use ⬤ to show 10. Write the number. Use ⚪ show eight. Write the number. Use red to draw counters to show a group of 10. Use yellow to draw counters to show eight more. What number does 10 and eight more make? Write the number.

1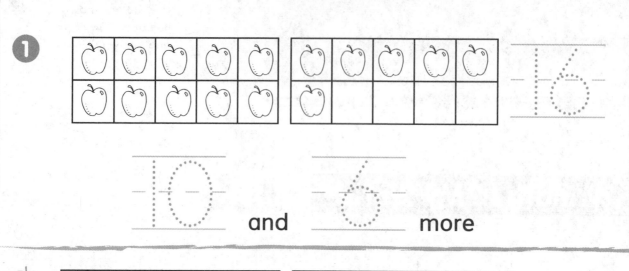

10 and 6 more

2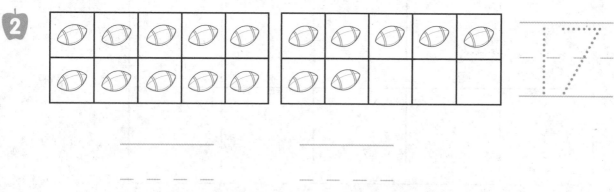

_____ and _____ more

3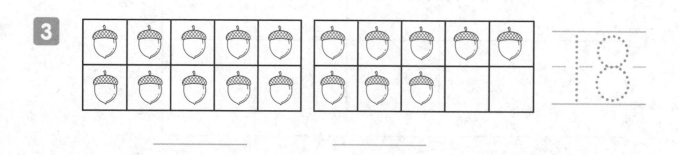

_____ and _____ more

 Directions: 1–3. Count 10. Use Work Mat 4 and red counters to show 10. Color the objects red to show 10. Write the number. Count how many more. Use yellow counters to show how many more. Color the other objects yellow to show how many more. Write the number. Trace the number made.

Independent Practice

 4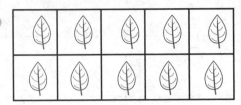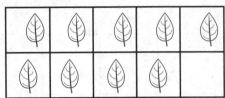

_____

_____ and _____ more

5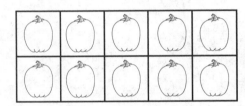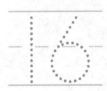

_____

_____ and _____ more

6

_____

_____ and _____ more

 Directions: 4–6. Count 10. Use Work Mat 4 and red counters to show 10. Color the objects red to show 10. Write the number. Count how many more. Use yellow counters to show how many more. Color the other objects yellow to show how many more. Write the number. Trace the number made.

7

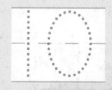 and 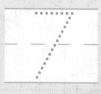 is  .

8

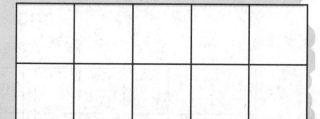

 and is  .

 Directions: 7–8. Trace 10. Draw objects to show 10 in the ten-frame. Color the objects red. Trace the other number. Draw objects to show that many in the other ten-frame. Color the objects yellow. Write the number made. In Exercise 7, explain to a friend how you could change the ten-frames to make 18.

Name ....

My Homework

Homework Helper

Need help? connectED.mcgraw-hill.com

1 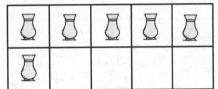 **16**

10 and **6** more

2 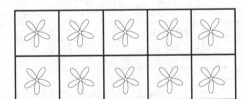 **19**

_____ and _____ more

 Directions: 1–2. Count 10. Color the objects red to show 10. Write the number. Count how many more. Color the other objects yellow to show how many more. Write the number. Trace the number made.

Copyright © McGraw-Hill Education

_____ and _____ more

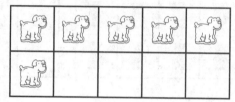

_____ and _____ more

_____ and _____ more

Directions: 3–5. Count 10. Color the objects red to show 10. Write the number. Count how many more. Color the other objects yellow to show how many more. Write the number. Trace the number made.

Math at Home Choose a number from 16 to 19. Guide your child to use objects to show the number as a group of 10 and a group of some more. Write the numbers.

Lesson 5
Take Apart Numbers 16 to 19

ESSENTIAL QUESTION ?
How do we show numbers 11 to 19 in another way?

Math in My World

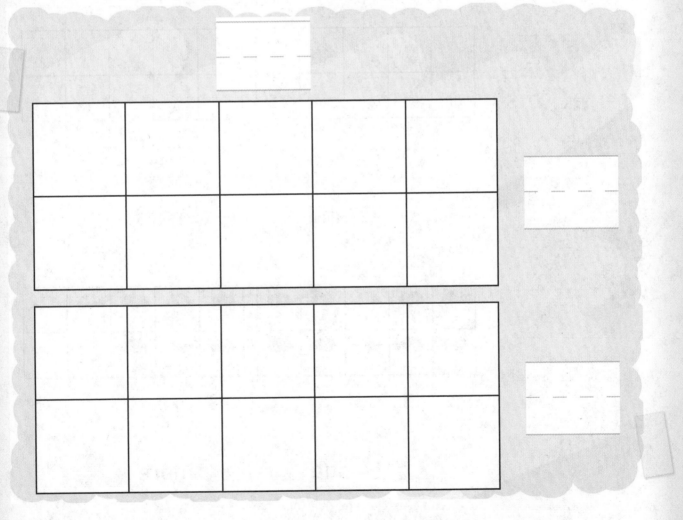

 Teacher Directions: Say a number between 16 and 19. Write the number above the ten-frames. Use to show the number. Use one ten-frame to show 10. Write the number 10. Use the other ten-frame to show the other counters. Write the number. Draw the counters in the ten-frames to show a group of 10 and some more.

Guided Practice

1 16

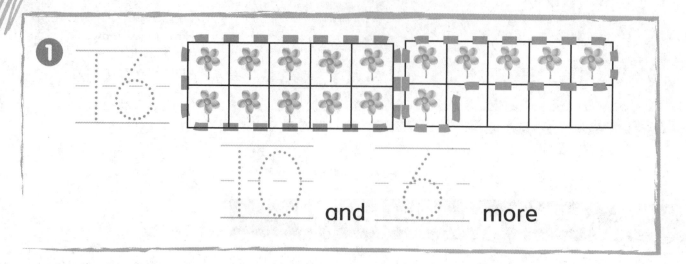

10 ___ and ___ 6 ___ more

2 17

_____ and _____ more

3 18

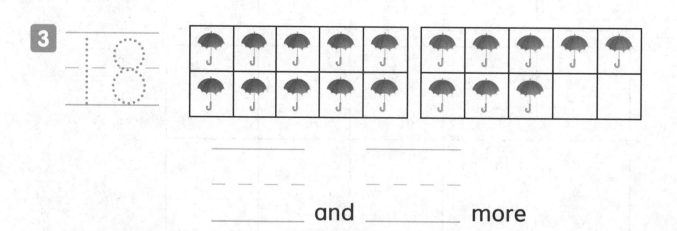

_____ and _____ more

 Directions: 1–3. Say the number. Trace it. Use Work Mat 4 and red counters to show the number. Take apart the number by showing 10 counters and some more. Write the numbers. Circle the group of objects that shows 10. Circle the group that shows some more.

Name

Independent Practice

 4 19

_____ and _____ more

 **5** 17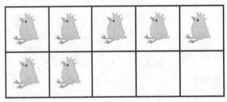

_____ and _____ more

 6 16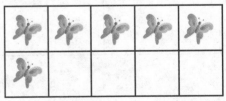

_____ and _____ more

Directions: 4–6. Say the number. Trace it. Use Work Mat 4 and red counters to show the number. Take apart the number by showing 10 counters and some more. Write the numbers. Circle the group of objects that shows 10. Circle the group that shows some more.

Copyright © McGraw-Hill Education

Brain Builders

7

_____ is _____ and _____ .

8

_____ is _____ and _____ .

 Directions: 7–8. Say the number. Trace the number. Draw objects to show the number. Circle the group of objects that shows 10. Circle the group that shows some more. Write the number sentence. Explain to a friend why it is best to take apart a number into a group of ten and some more.

Name _____

My Homework

Homework Helper

Need help? connectED.mcgraw-hill.com

1 17

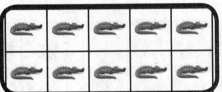 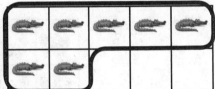

10 and 7 more

2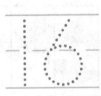

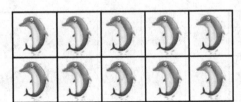

_____ and _____ more

 Directions: 1–2. Say the number. Trace it. Circle the group of objects that shows 10.
Write the number. Circle the group that shows some more. Write the number.

3 19

_____ _____

_____ and _____ more

4 17

_____ _____

_____ and _____ more

5 18

_____ _____

_____ and _____ more

Directions: 3–5. Say the number. Trace it. Circle the group of objects that shows 10. Write the number. Circle the group that shows some more. Write the number.

Math at Home Show your child 16 pieces of dry cereal. Have your child show the pieces in a group of 10 and a group of some more. Repeat with numbers 17, 18, and 19.

474   Chapter 7 • Lesson 5

My Review

Vocabulary Check

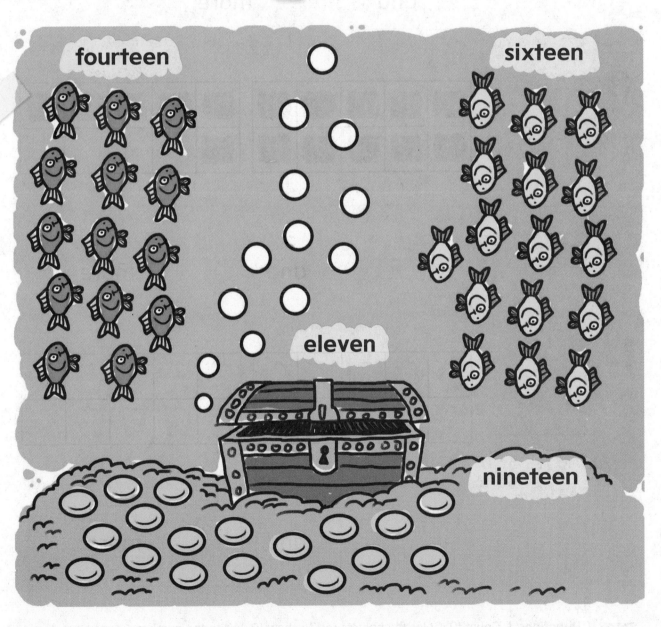

fourteen

sixteen

eleven

nineteen

 Directions: Count each group of fish. Circle the group of 16 fish. Put an X on the group of 14 fish. Count each coin. Say the number. Draw more to show 19. Color 11 bubbles.

Concept Check

1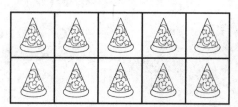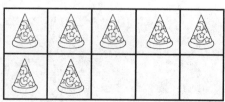

_____

_____ and _____ more

2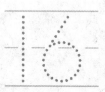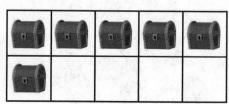

_____

_____ and _____ more

3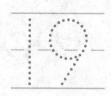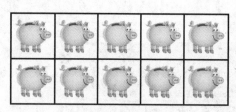

_____

_____ and _____ more

 Directions: 1. Count 10. Color the objects red to show 10. Write the number. Count how many more. Color the other objects yellow to show some more. Write the number. Trace the number made. **2–3.** Say the number. Trace it. Circle the group of objects that shows 10. Circle the group that shows some more. Write the numbers.

Brain Builders

4

1 5

_____ is _____ and _____ .

5

1 6

_____ is _____ and _____ .

Directions: 4–5. Say the number. Trace it. Draw objects in the ten-frames to show the number. Circle the group of objects that shows 10. Circle the group of objects that shows some more. Write the number sentence.

Reflect

Chapter 7

ESSENTIAL QUESTION
How do we show numbers 11 to 19 in another way?

1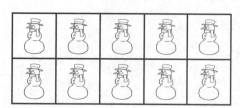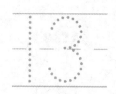

_____ and _____ more

2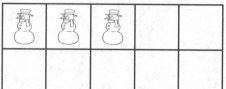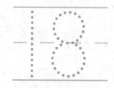

_____ and _____ more

3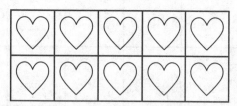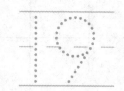

_____ and _____ more

 Directions: 1–3. Count 10. Use Work Mat 4 and red counters to show 10. Color the objects red to show 10. Write the number. Count how many more. Use yellow counters to show how many more. Color the other objects yellow to show some more. Write the number. Trace the number made.

Copyright © McGraw-Hill Education

Performance Task

Brain Builders

At the Pumpkin Patch

You can find fun things to do. You can pet zoo animals or go on a hay ride.

Show all your work to receive full credit.

Part A

Count 10. Color the objects orange to show 10. Write the number. Count the other objects. Color the other objects green to show how many more. Write the number. Trace the number made.

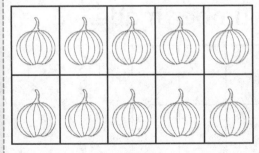 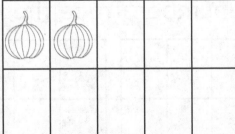 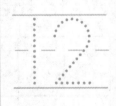

_____ and _____ more

Part B

Say the number. Trace it. Circle the group of chicks that shows 10. Write the number. Circle the group that shows some more. Write the number.

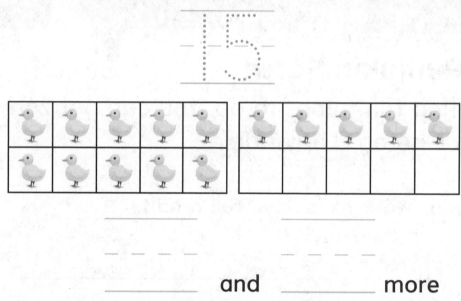

_____ and _____ more

Part C

Count 10. Color the leaves red to show 10. Write the number. Count how many more. Color the other leaves yellow to show how many more. Write the number. Trace the number made.

_____ and _____ more

Chapter

8 Measurement

ESSENTIAL QUESTION

How do I describe and compare objects by length, height, and weight?

Let's Get Camping!

Watch a video!

Watch

Name _____

Chapter 8 Project

Measuring Me

1 Look at the objects. Compare different objects by length, height, and weight to you.

2 Complete the sentences below.

Length

I am longer than a _____ .

I am shorter than a _____ .

Height

I am taller than a _____ .

I am shorter than a _____ .

Weight

I am heavier than a _____ .

I am lighter than a _____ .

3 Compare and talk about your sentences with a classmate. Did they use different objects? Do you agree with their sentences?

Name _____

1

2

3

4

 Directions: 1–2. Circle the object that is different. **3.** Circle the object that is longer. **4.** Circle the animal that you could hold in your hand.

Name _____

My Math Words

Review Vocabulary

bigger   smaller

 Directions: Circle the bigger duck. Draw an X on the smaller ducks. Trace each word. Draw a bigger frog next to the frog.

capacity

heavier

↑ heavier

height

holds less

holds less

holds more

holds more

length

Teacher Directions:
Ideas for Use

- Have students choose two objects in the classroom. Have students choose a vocabulary card and use the vocabulary word to compare the two objects.

- Have students name the letters in each word.

heavier

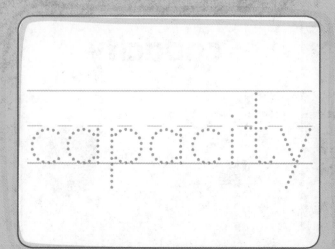

capacity

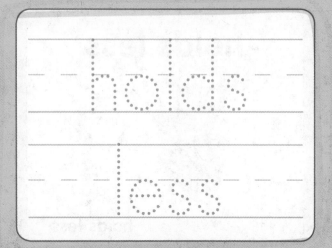

holds
less

height

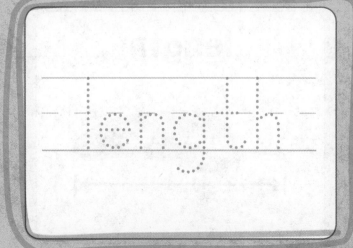

length

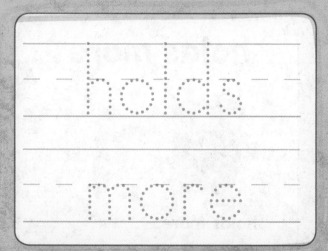

holds
more

My Vocabulary Cards

Vocab

Processes & Practices

lighter

lighter

longer

long

longer

shorter

short   shorter

taller

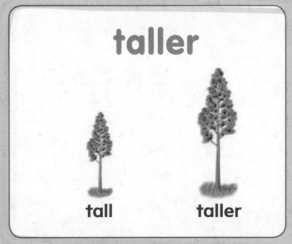

tall   taller

weight

Teacher Directions:
More Ideas for Use

- Tell students to create riddles for each word. Ask them to work with a friend to guess the word for each riddle.

- Help students use the blank card to write a word from a previous chapter that they would like to review.

longer

lighter

taller

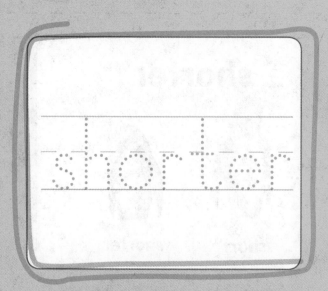

shorter

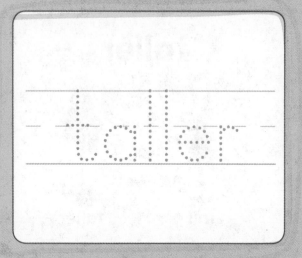

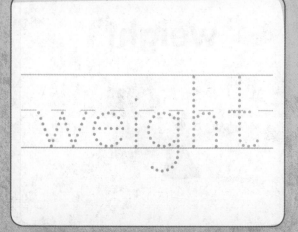

weight

My Foldable

 FOLDABLES® Follow the steps on the back to make your Foldable.

✂ — — — — — — — — — — — — — — — — —

 longer

shorter

shorter

taller

heavier

lighter

 Copyright © McGraw-Hill Education

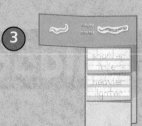
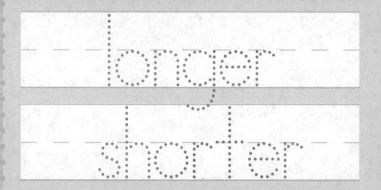
longer

shorter

shorter

taller

heavier

lighter

Name _____

ESSENTIAL QUESTION
How do I describe and compare objects by length, height, and weight?

Math in My World

Tools | Watch

 Teacher Directions: Use to make a train shorter than the fish. Draw the train above the fish. Use cubes to make a train longer than the fish. Draw the train below the fish.

1 length   longer   shorter

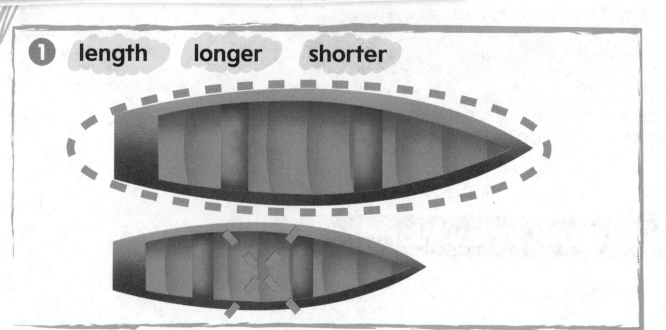

2

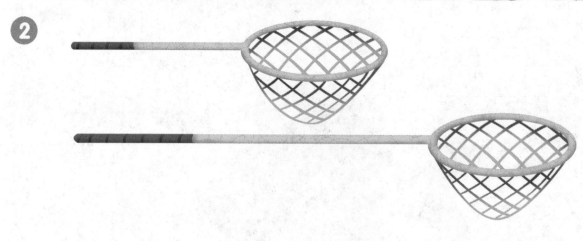

3

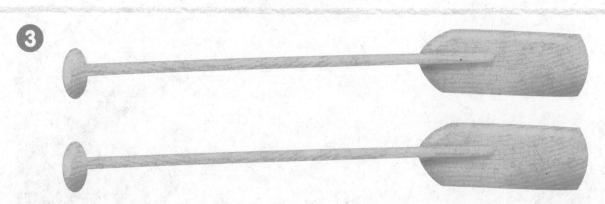

Directions: 1. Compare the objects. Trace the X on the object that is shorter. Trace the circle around the object that is longer. **2–3.** Compare the objects. Draw an X on the object that is shorter. Draw a circle around the object that is longer. If the objects are the same length, underline them.

Name

4

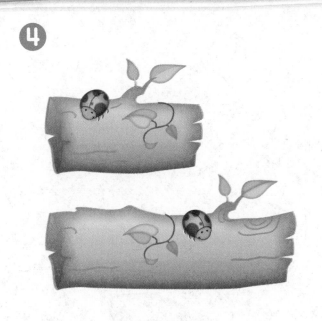

5

6

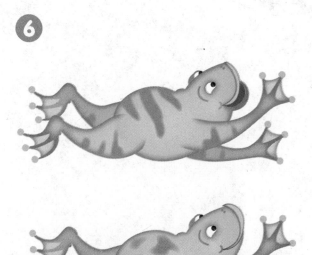

7

Directions: 4–7. Compare the objects. Draw an X on the object that is shorter. Draw a circle around the object that is longer. If the objects are the same length, underline them.

8

 Directions: 8. Use connecting cubes to make a train the same length as the snake. Use cubes to make a train shorter and a train longer than the snake. Draw a shorter snake with a green crayon above the snake and a longer snake with a green crayon below the snake. Use connecting cubes to make a cube train. Draw two snakes that are the same length as the cube train using a red crayon.

Name _____

My Homework

Homework Helper

Need help? connectED.mcgraw-hill.com

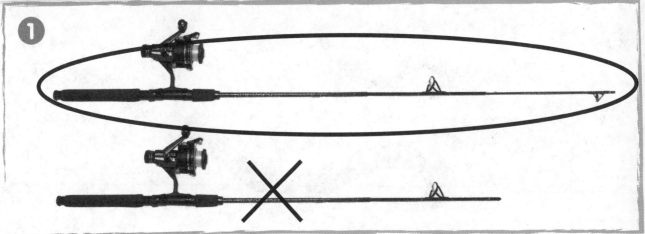

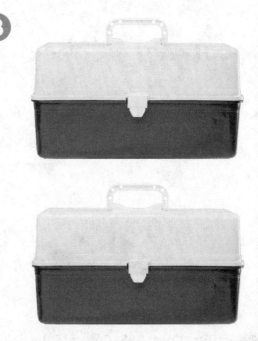

Directions: 1–3. Compare the objects. Draw an X on the object that is shorter. Draw a circle around the object that is longer. If the objects are the same length, underline them.

4

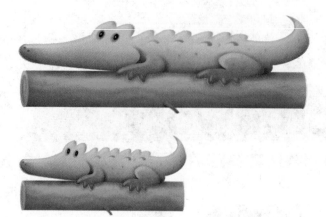

5

Vocabulary Check

6

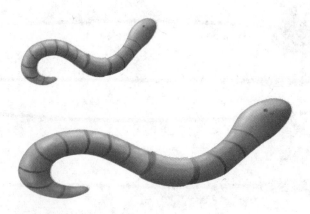

longer

shorter

 Directions: 4–5. Compare the objects. Draw an X on the object that is shorter.
Draw a circle around the object that is longer. **6.** Draw a line from the word *longer*
to the longer animal. Draw a line from the word *shorter* to the shorter animal.

Math at Home   Place a spoon and pencil on the table. Have your child tell which is
shorter, which is longer, or if the objects are the same length.

Name _____

Lesson 2
Compare Height

ESSENTIAL QUESTION ❓
How do I describe and compare objects by length, height, and weight?

Math in My World Tools Watch

 Teacher Directions: Use ▪️ to show a rocket that is taller than the rocket on the page. Trace the connecting cubes.

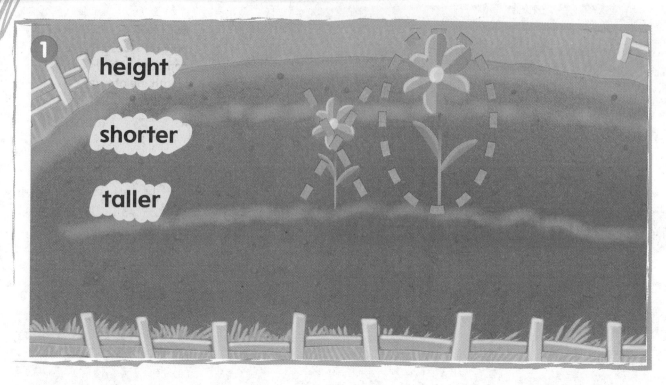

1 height

shorter

taller

2

3

Directions: I. Compare the objects. Trace the X on the object that is shorter. Trace the circle around the object that is taller. **2–3.** Compare the objects. Draw an X on the object that is shorter. Draw a circle around the object that is taller. If the objects are the same height, underline them.

Name _____

4

5

6

7

MY GARDEN

MY GARDEN

Directions: 4–7. Compare the objects. Draw an X on the object that is shorter. Draw a circle around the object that is taller. If the objects are the same height, underline them.

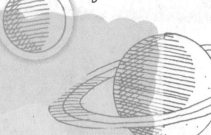

8

Directions: 8. Draw a tall object. Draw a short object beside it. Color the tall object red and the short object yellow. Explain to a classmate what the words short and tall mean.

Name _____

Homework Helper

Need help? connectED.mcgraw-hill.com

1

2

3

Directions: 1–3. Compare the objects. Draw an X on the object that is shorter. Draw a circle around the object that is taller. If the objects are the same height, underline them.

4

5

Vocabulary Check

6

taller   **shorter**

 Directions: 4–5. Compare the objects. Draw an X on the object that is shorter. Draw a circle around the object that is taller. **6.** Draw a line from the word *taller* to the taller animal. Draw a line from the word *shorter* to the shorter animal.

Math at Home   Gather a pencil and a crayon. Stand them side by side. Ask your child to tell which is shorter, which is taller, or if they are the same height. Compare other objects.

Copyright © McGraw-Hill Education   (t)Mark Steinmetz/McGraw-Hill Education, (c)Images-USA/USA/Alamy, (b)Kuttelvaserova Stuchelova/Shutterstock.com, (br)Yuri Kevhiev/Alamy

Name

How long?

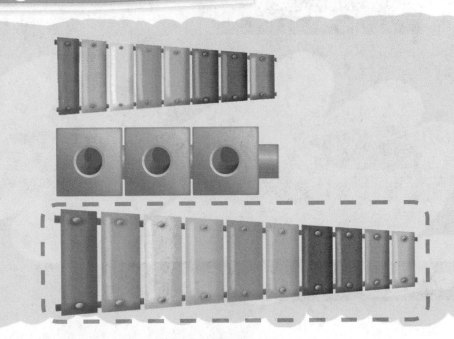

Guess, Check, and Revise

guess   check

 Teacher Directions: Compare the objects. Trace the circle around the object that is longer. Then guess how many cubes long the longer object is. Write the number. Use cubes to check. Is your answer close? Trace the cubes. Trace the number.

Guess, Check, and Revise

_____ _____
- - - - - - - - -
guess   check

Directions: Compare the objects. Circle the object that is longer. Then guess how many cubes long the longer object is. Write the number. Use cubes to check. Is your answer close? Trace the cubes. Write the number.

Name ....

How long?

Guess, Check, and Revise

_ _ _ _ _ _ _ _ _ _ _ _

_ _ _ _ _ _ _ _ _ _ _ _

guess   check

 Directions: Compare the objects. Circle the object that is longer. Then guess how many cubes long the longer object is. Write the number. Use cubes to check. Is your answer close? Trace the cubes. Write the number. Explain to a friend how you made your guess.

How long?

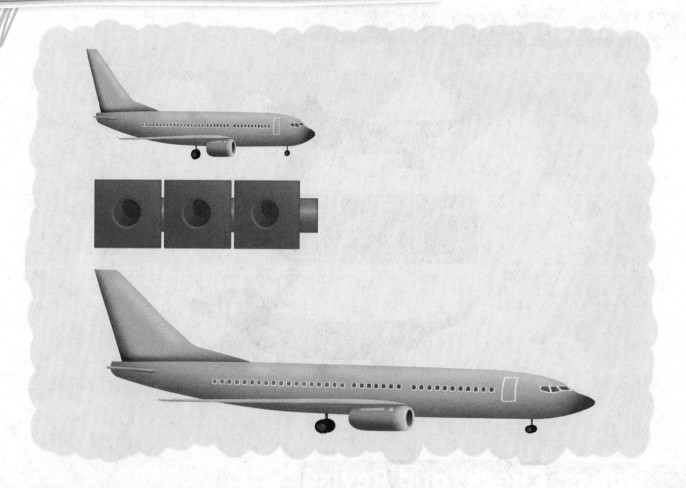

Guess, Check, and Revise

_____ _____

guess   check

 Directions: Compare the objects. Circle the object that is longer. Then guess how many cubes long the longer object is. Write the number. Use cubes to check. Is your answer close? Trace the cubes. Write the number.

Name _____

My Homework

How long?

ADMISSION TICKET

Admit One

ADMISSION TICKET

Admit One

Guess, Check, and Revise

guess   check

Directions: Compare the objects. Circle the object that is longer. Then guess how many pennies long the longer object is. Write the number. Use pennies to check. Is your answer close? Trace the pennies. Trace the number.

How long?

Guess, Check, and Revise

_____ _____

guess   check

Directions: Compare the objects. Circle the object that is longer. Then guess how many pennies long the longer object is. Write the number. Use pennies to check. Is your answer close? Trace the pennies. Write the number.

Math at Home Take advantage of problem-solving opportunities during daily routines such as riding in the car, bedtime, doing laundry, putting away groceries, and so on.

Name _____

Vocabulary Check

1 **longer**

shorter

2

shorter   **taller**

Concept Check

3

 Directions: 1. Look at the pencil. Draw a pencil that is shorter. **2.** Look at the cup. Draw a cup that is taller. **3.** Compare the objects. Draw an X on the object that is shorter. Draw a circle around the object that is taller.

Directions: 4–5. Compare the objects. Draw an X on the object that is shorter. Draw a circle around the object that is taller. **6–7.** Compare the objects. Draw an X on the object that is shorter. Draw a circle around the object that is longer. If the objects are the same length, underline them.

Name

ESSENTIAL QUESTION
How do I describe and compare objects by length, height, and weight?

 Math in My World Watch ▶

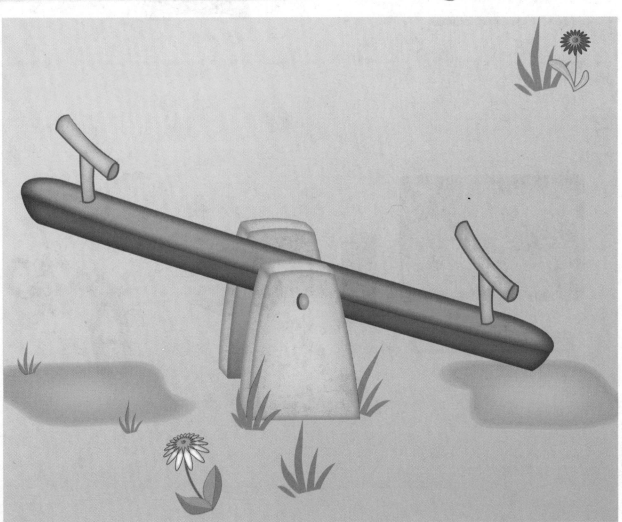

🦉 **Teacher Directions:** Find two different objects in the classroom. Tell which object is heavy and which is light. Draw a picture of each object on the teeter-totter to show which object is heavy and which is light.

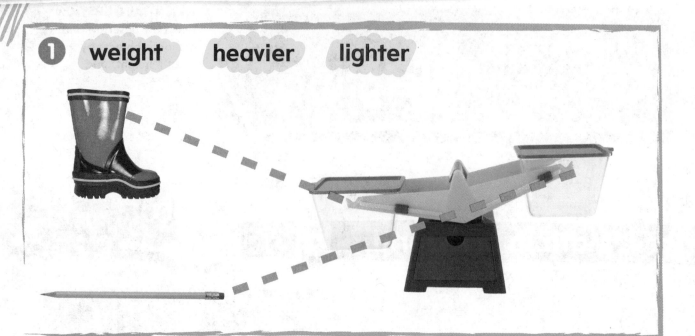

① weight   heavier   lighter

②

③

Directions: I. Compare the objects. Trace the line from each object to the place on the balance scale that shows its weight. **2–3.** Compare the objects. Circle the heavier object. Draw an X on the lighter object. If the objects weigh the same, underline them.

Independent Practice

4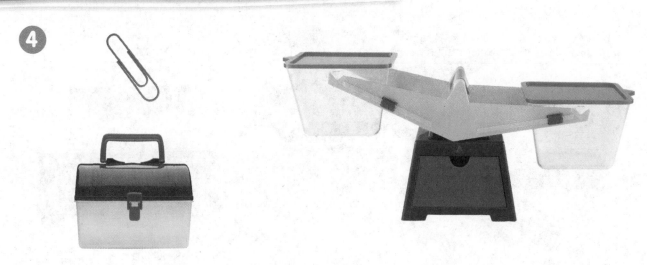

5

6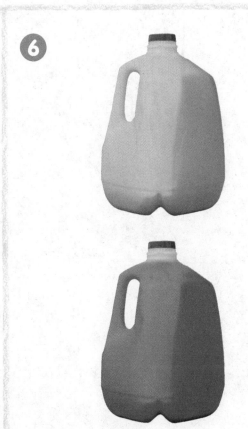

 Directions: 4. Compare the objects. Draw a line from each object to the place on the balance scale that shows its weight. **5–6.** Compare the objects. Circle the heavier object. Draw an X on the lighter object. If the objects weigh the same, underline them.

Brain Builders

7

Directions: 7. Draw an X on items that are too heavy to lift. Circle the items that are light enough to carry. Tell a friend as many items that you can think of that are too heavy to lift. Then tell a friend as many items that you can think of that are light enough to carry.

Name

My Homework

Homework Helper

Need help? connectED.mcgraw-hill.com

1

2

3

 Directions: 1–3. Compare the objects. Circle the heavier object. Draw an X on the lighter object. If the objects weigh the same, underline them.

4

5

Vocabulary Check

6 lighter

heavier

 Directions: 4–5. Compare the objects. Circle the heavier object. Draw an X on the lighter object. If the objects weigh the same, underline them. **6.** Draw a line from the word *lighter* to the lighter object. Draw a line from the word *heavier* to the heavier object.

Math at Home Use a canned good and an empty cup. Place one item in each of your child's hands. Ask your child which hand holds the heavier item and which hand holds the lighter item.

Name _____

ESSENTIAL QUESTION
How do I describe and compare objects by length, height, and weight?

Math in My World

 Teacher Directions: Choose two real-world objects. Draw them. Describe the length of each object. Circle the object that is longer. Describe the weight of each object. Draw an X on the object that weighs more.

Guided Practice

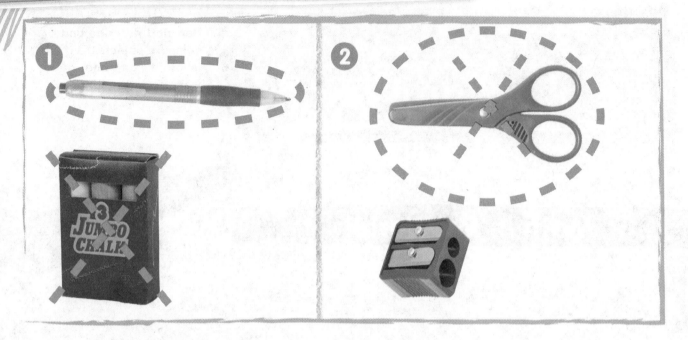

 Directions: 1–2. Look at the objects. Describe the length of each object. Circle the object that is longer. Describe the weight of each object. Draw an X on the object that weighs more. **3–4.** Look at the objects. Describe the height of each object. Circle the object that is taller. Describe the length of each object. Draw an X on the object that is longer.

Name _____

5

6

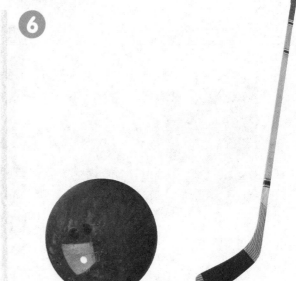

7

8

 Directions: 5–8. Look at the objects. Describe the height of each object. Circle the object that is taller. Describe the weight of each object. Draw an X on the object that weighs more.

9

10

Directions: 9. Look at the box of bandages. Describe the objects in the row. Circle the object that is longer and weighs more than the box of bandages. **10.** Look at the bowl. Describe the objects in the row. Circle the object that is taller and weighs more than the bowl. Tell a friend another item that is taller and weighs more than the bowl. Then tell a friend a different item that is shorter and weighs less than the bowl.

Name _____

My Homework

Homework Helper

Need help? connectED.mcgraw-hill.com

1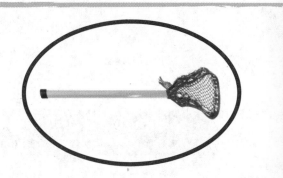

2

3

 Directions: 1. Look at the objects. Describe the length of each object. Circle the object that is longer. Describe the weight of each object. Draw an X on the object that weighs more. **2–3.** Look at the objects. Describe the height of each object. Circle the object that is taller. Describe the length of each object. Draw an X on the object that is longer.

4

5

6

 Directions: 4. Look at the objects. Describe the height of each object. Circle the object that is taller. Describe the weight of each object. Draw an X on the object that weighs more. **5–6.** Look at the objects. Describe the length of each object. Circle the object that is longer. Describe the weight of each object. Draw an X on the object that weighs more.

Math at Home Gather two items in your house. Ask your child to tell which is taller. Ask your child to tell which is longer. Ask your child to tell which item weighs more.

Name

Compare Capacity

 Math in My World Tools Watch

 Teacher Directions: Use ▇ to fill in both sandboxes on the page. Tell which sandbox holds more. Circle the sandbox that holds less.

1

capacity

holds more

holds less

2

3

Directions: 1. Compare the containers. Trace the circle around the object that holds more. Trace the X on the object that holds less. **2–3.** Compare the containers. Circle the object that holds more. Draw an X on the object that holds less. If the objects hold the same, underline them.

Name

4

5

6

 Directions: 4–6. Compare the containers. Circle the object that holds more. Draw an X on the object that holds less. If the objects hold the same, underline them.

Brain Builders

7

 Directions: 7. Look at the object in the box. Draw an X on items that hold more than the object. Circle the items that hold less than the object. Explain to a friend how you know if one object holds more than another object.

Name _____

My Homework

Homework Helper

Need help? connectED.mcgraw-hill.com

1

2

3

 Directions: 1–3. Compare the containers. Circle the object that holds more. Draw an X on the object that holds less. If the objects hold the same, underline them.

4

5

Vocabulary Check

6   holds more   holds less

Directions: 4–5. Compare the containers. Circle the object that holds more. Draw an X on the object that holds less. If the objects hold the same, underline them. **6.** Draw a line from the object that holds more to the words *holds more*. Draw a line from the object that holds less to the words *holds less*.

 Math at Home   Use a sauce pan and empty cup. Ask your child which object holds more and which object holds less.

Name _____

My Review

Vocabulary Check

taller

shorter

holds less

holds more

heavier

lighter

 Directions: 1. Circle the tent that is taller. Draw an X on the tent that is shorter.
2. Circle the cooler that holds more. Draw an X on the cooler that holds less.
3. Circle the bag that is heavier. Draw an X on the bag that is lighter.

Concept Check

1

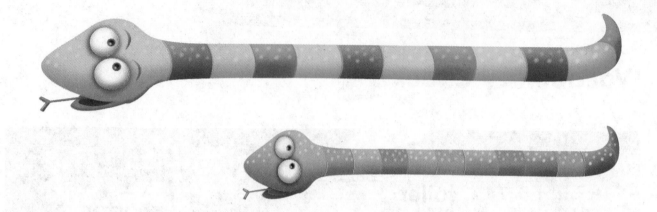

2

3

 Directions: 1. Compare the objects. Draw an X on the object that is shorter. Circle the object that is longer. **2.** Draw an X on the object that is heavier. Circle the object that is lighter. **3.** Draw an X on the object that holds more. Circle the object that holds less.

Brain Builders

4

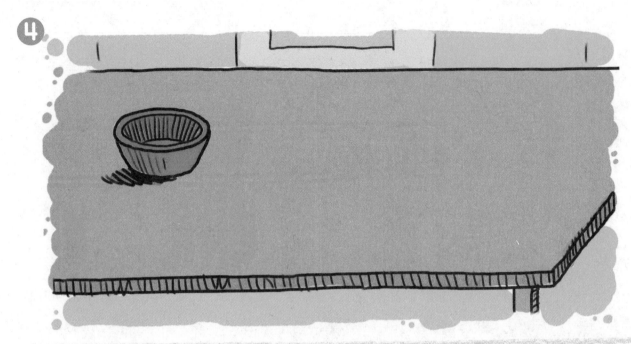

5

Directions: 4. Look at the bowl. Draw a bowl that holds more next to it.
5. Look at the tent. Draw a shorter tent next to it.

Chapter 8

ESSENTIAL QUESTION
How do I describe and compare objects by length, height, and weight?

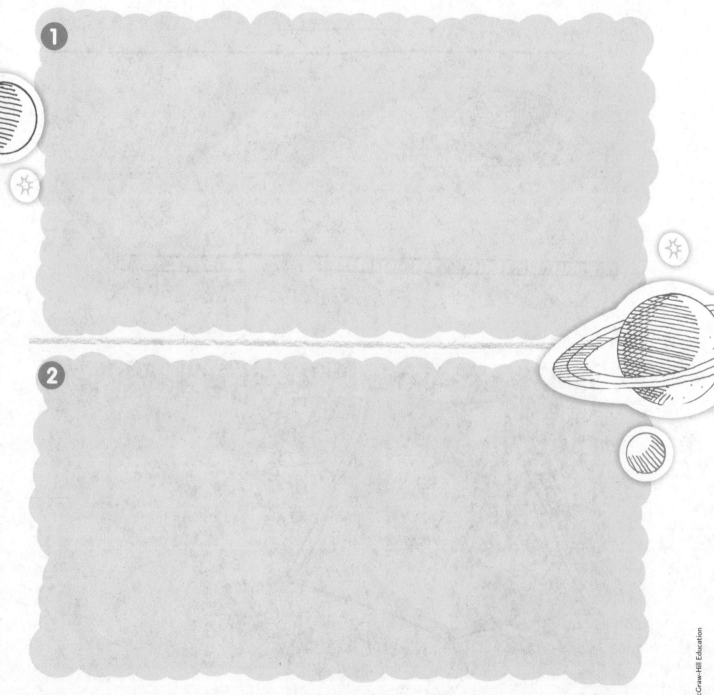

1

2

Directions: 1. Draw one long object. Draw a shorter object below it. **2.** Draw a tall object. Draw a shorter object beside it. Have a classmate tell which object is short and which object is tall.

Performance Task

Brain Builders

Let's Go to the Country

In the country you can visit a farm and see different kinds of animals and food grown there.

Show all your work to receive full credit.

Part A

Compare the objects. Circle the object that holds less. Draw an X on the object that holds more.

Part B

Draw an X on the turtle that is longer.

Part C

Compare the objects. Circle the objects that are light enough to lift by yourself.

Part D

Compare the objects. Circle the object that is taller. Draw an X on the object that is shorter.

Chapter 9 Classify Objects

Earth Needs Our Help!

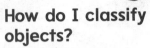

WE RECYCLE

Watch

Watch a video!

Name _____

Chapter 9 Project

Sort Classroom Objects

1 Choose a way to sort the objects and then sort them.

2 Record the way you sorted the objects below.

We sorted by _____ .

We sorted by _____ .

We sorted by _____ .

We sorted by _____ .

We sorted by _____ .

3 Choose another way to sort the objects and sort them again. Record the way you sorted the objects above. Try to find as many ways to sort the objects as you can. Record all the ways.

Name _____

Am I Ready?

1

2

3

4

 Directions: 1. Trace the dashed marks. **2.** Circle the car. Draw an X on the tree.
3. Color the truck red. Color the ball blue. Color the bird yellow.
Color the flower purple. **4.** Circle the object that is small.

Chapter 9   533

My Math Words

Vocab

Review Vocabulary

Directions: Trace each word. Draw lines to match each word with the recycling bins that show that size.

My Vocabulary Cards

alike

different

shape

size

sort

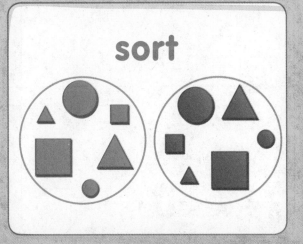

Teacher Directions:
Ideas for Use

- Have students choose two words. Have them tell if any of the letters in the words are alike. Have them point to letters that are different.

- Have students name the letters in each word.
- Have students use the blank card to write their own vocabulary word.

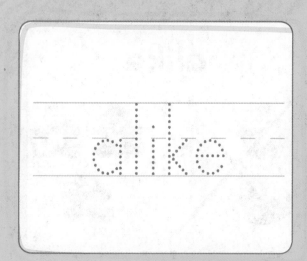

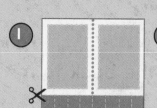

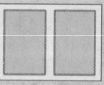

Name

Alike and Different

Math in My World

Tools Watch

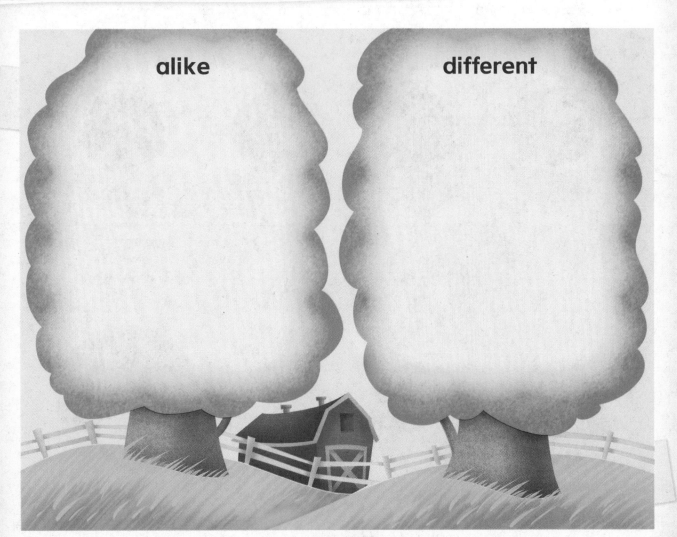

alike   **different**

 Teacher Directions: Place some along the grass on the page. Describe the blocks using the words *alike* and *different*. Place the pattern blocks that are the same on the tree with the word *alike*. Place the other pattern blocks on the tree with the word *different*. Trace and color the blocks.

1 alike   different

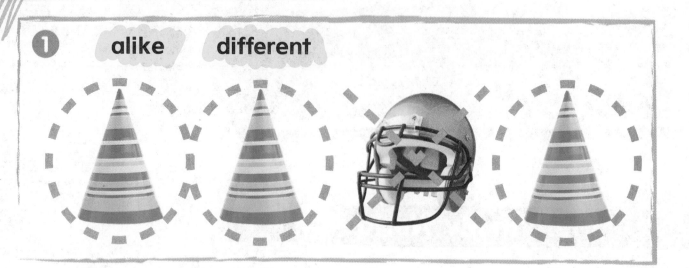

2

3

Directions: I. Look at the objects. Trace each circle to show the objects that are alike. Trace the X to show the object that is different. Tell why it does not belong.
2–3. Look at the objects. Circle the objects that are alike. Draw an X on the one that is different. Tell why it does not belong.

Independent Practice

4

5

6

 Directions: 4–6. Look at the objects. Circle the objects that are alike. Draw an X on the one that is different. Tell why it does not belong.

Directions: 7. Look at each group of objects near each animal. Circle the objects that are alike. Draw an X on each object that is different. Explain to a friend why it does not belong. Point to the shirt that is different. Draw the missing parts to make it the same as the other shirts.

Name _____

My Homework

Homework Helper

eHelp

Need help? connectED.mcgraw-hill.com

1

2

3

Directions: 1–3. Look at the objects. Circle the objects that are alike. Draw an X on the one that is different. Tell why it does not belong.

4

Vocabulary Check

5 alike

6 different

 Directions: 4. Look at the objects. Circle the objects that are alike. Draw an X on the one that is different. Tell why it does not belong. **5.** Look at the objects. Circle the objects that are alike. **6.** Look at the objects. Draw an X on the one that is different.

Math at Home Draw a picture of some items in each room of your home. Ask your child to tell how the items are alike or different.

Name

Problem Solving

STRATEGY: Use Logical Reasoning

ESSENTIAL QUESTION
How do I classify objects?

What does not belong?

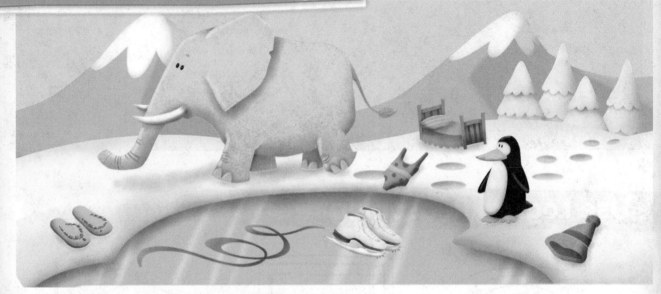

Use Logical Reasoning

 Teacher Directions: Look at the objects below the picture. Color the objects that do not belong outside in the cold. Explain your answer.

What does not belong?

Use Logical Reasoning

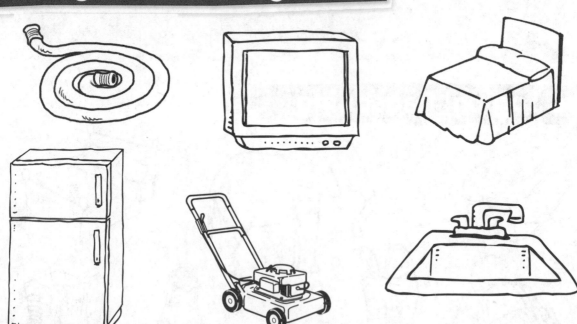

Directions: Look at the objects below the picture. Color the objects that do not belong in the yard. Explain your answer.

What does not belong?

Use Logical Reasoning

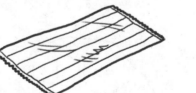

Directions: Look at the objects below the picture. Color the objects that do not belong on the beach. Explain your answer to a friend.

Use Logical Reasoning

Directions: Look at the objects below the picture. Color the objects that are not classroom supplies. Explain your answer.

Name

My Homework

What does not belong?

Use Logical Reasoning

Directions: Look at the objects below the picture. Color the objects that do not belong on a soccer field. Explain your answer.

What does not belong?

Use Logical Reasoning

Directions: Look at the objects below the picture. Color the objects that do not belong in the park. Explain your answer.

Math at Home Take advantage of problem-solving opportunities during daily routines, such as putting away groceries. Ask what belongs in the cupboard or in the freezer.

Lesson 3
Sort by Size

Math in My World
Tools Watch

Teacher Directions: Use six ▲ of the same color. Use two large buttons and four small buttons. Sort the buttons by size. Place the sorted buttons in the treasure box. Trace and color them to show how you sorted.

1 sort size

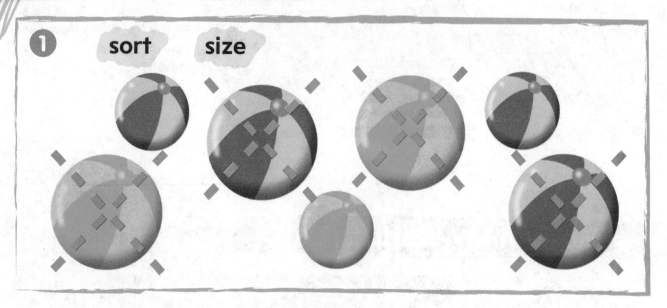

2

3

 Directions: I. The beach balls were sorted by size. Trace the Xs to show how the beach balls were sorted. **2–3.** Sort the objects by size. Draw Xs to show how you sorted.

Name _____

4

5

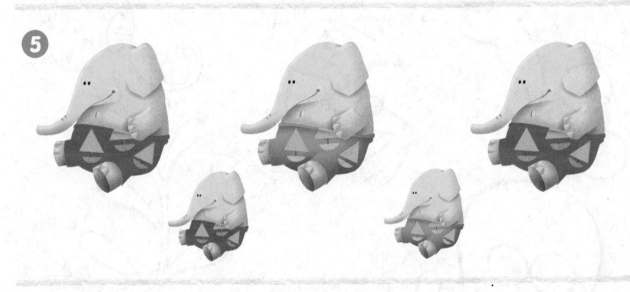

6

 Directions: 4–6. Sort the objects by size. Draw Xs to show how you sorted.

 Directions: 7. Look at the pairs of objects. Describe their sizes. Sort the objects by size. Color the large objects purple and the small objects yellow. Tell a friend how you could sort the animals by how many legs they have.

My Homework

Homework Helper   eHelp

Need help? connectED.mcgraw-hill.com

1

2

3

 Directions: 1-3. Sort the objects by size. Draw Xs to show how you sorted. Tell how you sorted.

4

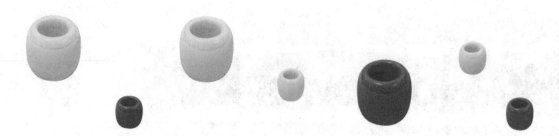

Vocabulary Check

5 sort

6 size

Directions: 4. Sort the objects by size. Draw Xs to show how you sorted. Tell how you sorted. **5.** Sort the recycle bins by size. Draw Xs on the large recycle bins. **6.** Draw Xs on the objects that are the same size.

Math at Home Place kitchen towels and wash cloths in a group. Have your child sort the items by size.

Name

Vocabulary Check

1 alike

2 sort

Concept Check

3

 Directions: 1. Circle the flowers that are alike. **2.** Sort the pillows by size. Draw Xs on the pillows to show how you sorted. **3.** Circle the objects that are alike. Draw an X on the one that is different.

Chapter 9   557

4

5

6

 Directions: 4–5. Circle the objects that are alike. Draw an X on the one that is different. **6.** Sort the bugs by size. Draw Xs to show how you sorted.

Name
....

Lesson 4
Sort by Shape

ESSENTIAL QUESTION **?**
How do I classify objects?

 Math in My World   Tools   Watch

 Teacher Directions: Use red ■ of two different shapes. Place the attribute blocks in the red box. Sort the blocks by shape into groups. Place one group in the other box. Trace the shapes to show how you sorted.

1 shape

2

3

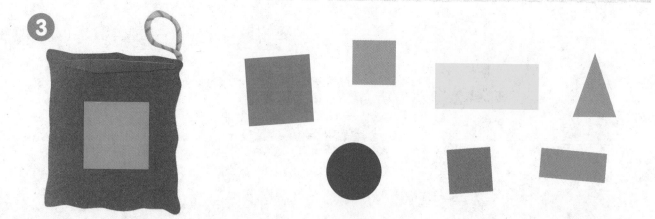

Directions: 1. Look at the shape on the bag. Sort the shapes in the group. Trace the circles around the shapes that belong in the bag. **2–3.** Look at the shape on the bag. Sort the shapes in the group. Draw a circle around the shapes that belong in the bag. Tell how you decided.

Independent Practice

What's in the bag?

④

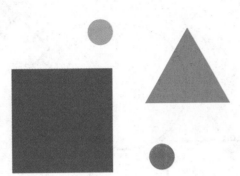

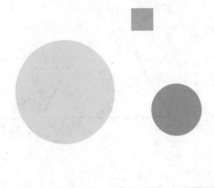

⑤

⑥

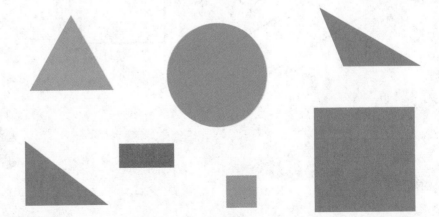

Directions: 4–6. Look at the shape on the bag. Sort the shapes in the group.
Draw a circle around the shapes that belong in the bag. Tell how you decided.

7

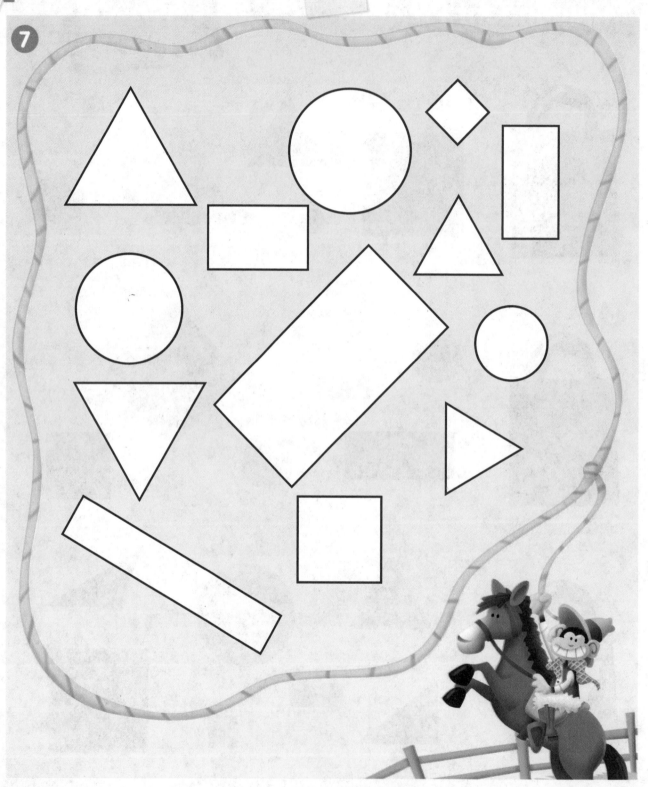

Directions: 7. Sort the shapes. Color all ⃝ red, all ▭ orange, all ☐ green, and all △ blue. Explain to a friend how you sorted the shapes.

My Homework

Homework Helper eHelp

Need help? connectED.mcgraw-hill.com

1
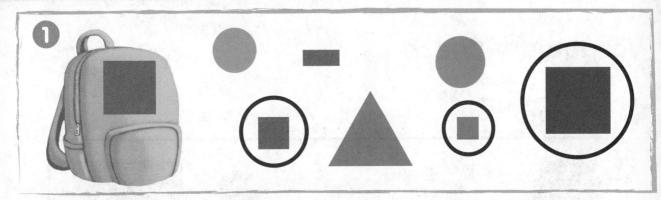

2

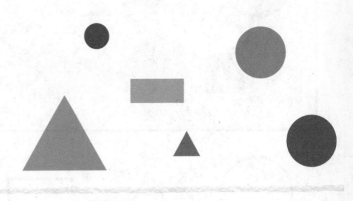

3

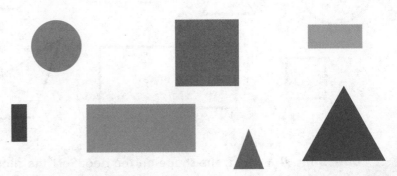

Directions: 1–3. Look at the shape on the bag. Sort the shapes in the group. Draw a circle around the shapes that belong in the bag.

4

Vocabulary Check

5 **shape**

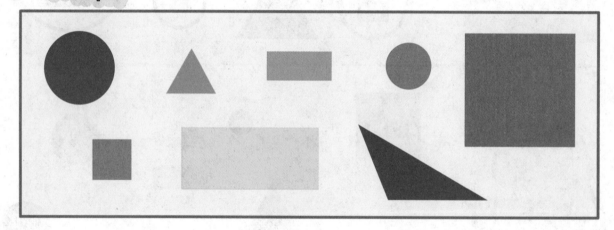

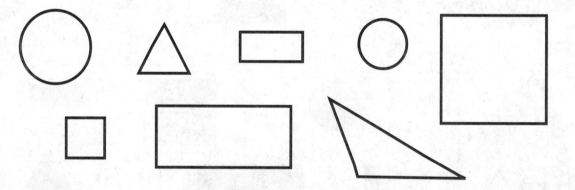

Directions: 4. Look at the shape on the bag. Sort the shapes in the group. Draw a circle around the shapes that belong in the bag. **5.** Look at the shapes in the red box. Color the shapes below to match.

Math at Home Mix round and square shaped cereal in a bowl. Have your child sort the cereal by shape into two different groups.

Name ....

 Math in My World `Tools`

 Teacher Directions: Use small attribute buttons. Place four ▦, three ●, and three ▲ randomly on the plate. Sort the buttons by shape into groups. Sort the groups by count. Tell how you sorted. Trace the buttons to show how you sorted.

1

2

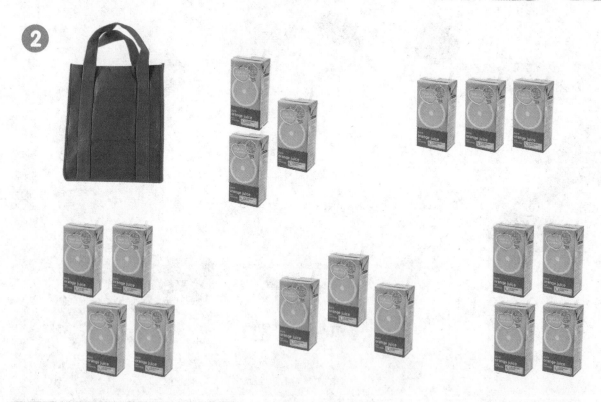

Directions: 1. Count the bottles in each group. Trace the Xs to show how the bottles are sorted. **2.** Count the juice boxes in each group. Sort the juice boxes by count. Draw Xs to show how you sorted.

Name

3

4

Directions: 3. Count the lightbulbs in each group. Sort the lightbulbs by count. Draw Xs to show how you sorted. **4.** Count the bottles in each group. Sort the bottles by count. Draw Xs to show how you sorted.

5

 Directions: 5. Count the objects in each group of food. Sort the objects by count. Draw Xs to show how you sorted. Sort by count again and circle the groups to show how you sorted. Explain to a friend why no other groups are sorted with the fruit cups.

Name

_____

My Homework

Homework Helper eHelp

Need help? connectED.mcgraw-hill.com

 Directions: 1. Count the dogs. Sort the dogs by count. Draw Xs to show how you sorted. **2.** Count the fish. Sort the fish by count. Draw Xs to show how you sorted.

3

4

Directions: 3. Count the hamsters. Sort the hamsters by count. Draw Xs to show how you sorted. **4.** Count the cats. Sort the cats by count. Draw Xs to show how you sorted.

Math at Home Use buttons or another small object. Show a group of three, a group of four, a group of three, a group of four, and another group of four. Have your child sort the groups by count.

Name _____

My Review

Vocabulary Check

 Directions: 1. Look at the attribute buttons. Point to a shape. Draw an X on the buttons that are the same shape. **2.** Look at the blocks. Sort the blocks by size. Circle the blocks to show how you sorted.

Concept Check

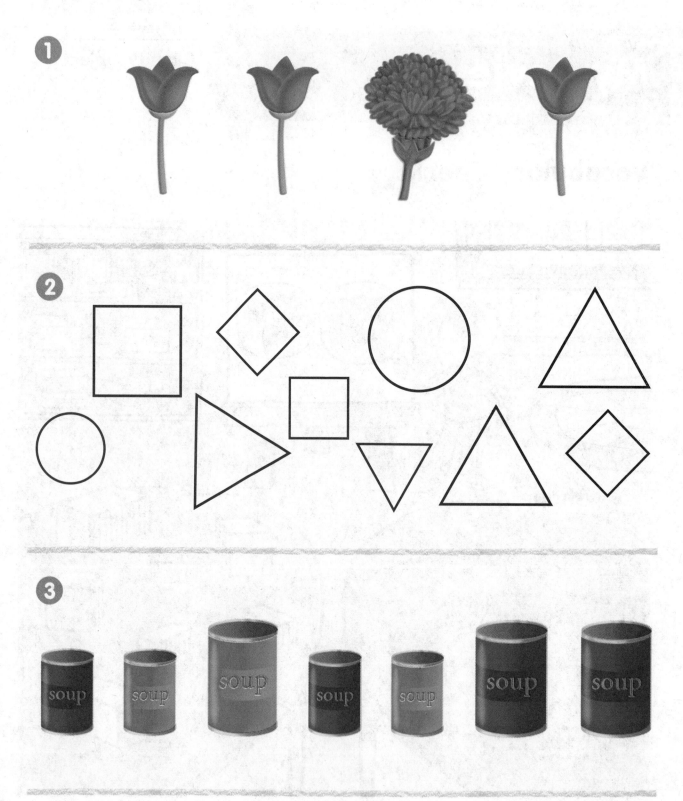

 Directions: I. Circle the objects that are alike. Draw an X on the one that is different. **2.** Sort the shapes. Color all △ blue, all ▢ red, and all ◯ green. **3.** Sort the cans by size. Draw Xs on the cans to show how you sorted.

Name

Brain Builders

Directions: 4. Look at the objects. Sort the objects by size. Draw lines from the objects to the recycle bins to show how you sorted.

shape   size

 Directions: Place attribute buttons on a tree. Sort the buttons by size or shape. Move the sorted buttons to the other tree. Trace the buttons on this tree to show how you sorted. Circle the word that tells how you sorted.

Performance Task

Brain Builders

A Visit to the Garden Center

At a garden center, you can see vegetables, flowers, trees, and bushes. Plants are everywhere and can be found in different colors, shapes, and sizes.

Show all of your work to receive full credit.

Part A

Circle the flowers that are alike. Draw an X on the flower that is different.

Part B

Draw Xs on the items that do not belong in a garden.

Part C

Sort by size. Underline the larger trees to show how you sorted.

Part D

Some items at the garden center look like the shapes below. Sort by shape. Circle all the shapes that are alike to show how you sorted.

Part E

Sort the groups by count. Circle the groups that have five to show how you sorted.

10 Position

We See Animals in Action!

Watch a video!

Watch ▶

Name

Chapter 10 Project

My Position Book

1. You will make a book to show the position words you learn in this chapter.

2. After you have learned a position word, draw a picture in your book to show the position word.

3. Your teacher will write the position word you learned on the board. Write that word above your picture.

4. When your book is finished, share your book with a partner. Explain to your partner each position that is shown in your book.

5. When your book is finished, draw a picture of you with the sun above you and grass below you. Draw a friend next to you and a tree behind you.

Name

Am I Ready?

1

2

3

Directions: 1. Draw eyes, a nose, and mouth on the face.
2. Color the bus wheels. **3.** Draw the missing chair legs.

My Math Words

Review Vocabulary

alike   different

Directions: Trace each word. Discuss how the birds are alike and how they are different. Color the birds that are alike blue. Draw Xs on the birds that are different. Explain your answer.

My Vocabulary Cards

Vocab
abc

Processes
& Practices

above

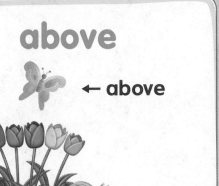

← above

behind

← behind

below

← below

beside

↑
beside

in front of

← in front of

next to

↑
next to

Teacher Directions:
Ideas for Use

- Ask students to arrange the cards to show the meaning of each word. For example, they might place the *above* card over the *below* card.

- Have students sort the words by the number of letters in each word.

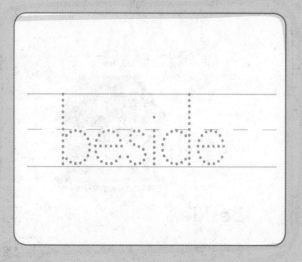

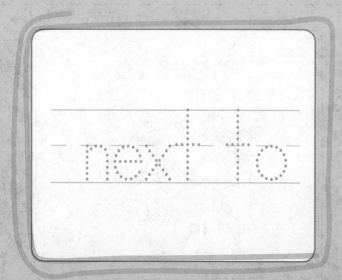

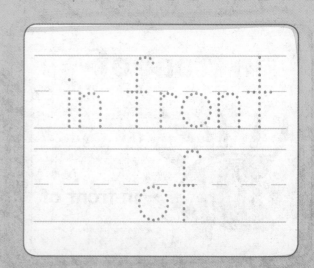

My Foldable

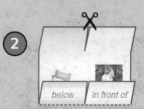
behind

above

in front of

below

Name _____

ESSENTIAL QUESTION
How do I identify positions?

 Math in My World Tools Watch

 Teacher Directions: Place a 🟦 on the page. Trace the cube. Use the words *above* or *over* and *below* or *under* to tell a partner where your cube is placed.

1 above below

2

3

Directions: 1. Describe the position of each monkey. Use *above* or *over* and *below* or *under*. Trace the circles around the monkeys that are above the branch. Trace the lines under the monkeys that are below the branch. **2.** Describe the position of each banana. Use *above* or *over* and *below* or *under*. Circle the bananas that are above the monkey. Underline the bananas that are below a monkey. **3.** Draw a banana below each monkey.

Independent Practice

4

5

6

 Directions: 4. Circle the fish that is above a dolphin. Underline the fish that are below a dolphin. **5.** Circle the crabs that are above a snake. Underline the crab that is below a snake. **6.** Describe the position of each fish. Use the words *above* and *below* or *over* and *under*. Circle the fish that is above. Draw an X on the fish that is below.

Brain Builders

Directions: 7. Draw a cloud above, or over, the airplane. Draw a rabbit below, or under, the grass. Draw a window above, or over, the door on the house. Circle the object that is above the house and below the sun. Explain to a friend how you know.

Name

My Homework

Homework Helper

Need help? connectED.mcgraw-hill.com

1

2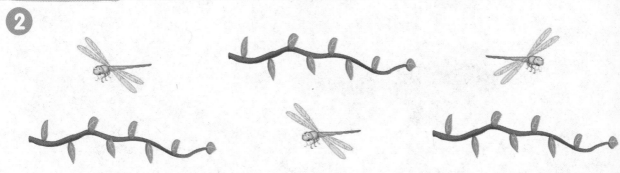

3

Directions: 1. Circle the bees that are above a flower. Underline the bee that is below a flower. **2.** Circle the dragonflies that are above a branch. Underline the dragonfly that is below a branch. **3.** Circle the butterflies that are above a net. Underline the butterfly that is below a net.

Chapter 10 • Lesson 1   587

④

Vocabulary Check

⑤ above

⑥ below

Directions: 4. Describe the position of each bird. Use *above* and *below*. Circle the bird that is above. Draw an X on the bird that is below. **5.** Draw a fish above the shark. **6.** Draw a seashell below the jellyfish.

Math at Home Give your child directions using the words *above* and *below*. For example, stack these books above the games on the shelf.

Name

Lesson 2
In Front of and Behind

ESSENTIAL QUESTION
How do I identify positions?

 Math in My World   Tools   Watch

 Teacher Directions: Draw a box around the object that is in front of the sand castle. Draw an X on each animal that is on the beach in front of the ladies. Draw an object behind the dolphin.

Directions: I. Trace the box around the object that is in front of the school. **2.** Trace the X on the object that is behind the plastic tube. **3.** Draw an X on each student that is in front of the bus. **4.** Draw a box around the raccoon that is in front of the slide.

Name

5

6

7

8

 Directions: 5. Draw a box around the animal that is behind the alligator. **6.** Draw an X on the animal that is in front of the cat. **7.** Draw a box around the animal that is behind the hippo. **8.** Draw an object. Draw another object behind your object.

Directions: 9. Draw an X on the girl that is in front of the house. Draw a box around the flower pot that is in front of the house. Draw a circle around the flower pot that is behind another flower pot. Circle the bear that is behind the raccoon. Explain to a friend how you know the bear is behind the raccoon.

My Homework

Homework Helper

eHelp

Need help? connectED.mcgraw-hill.com

1

2

3

 Directions: 1. Draw a box around the fish that is behind the treasure chest. **2.** Draw a box around the mouse that is behind the snake. **3.** Draw a box around the animal that is in front of the barn.

4

Vocabulary Check

5 in front of

6 behind

Directions: 4. Draw an X on the animal that is behind the fence. **5.** Draw an apple in front of the pig. **6.** Draw a flower behind the goat.

Math at Home Have your child look in the cupboard or refrigerator. Choose an item that has an item in front of it and behind it. Have your child tell which objects are in front of and behind. Choose another object and repeat.

Name

Check My Progress

Vocabulary Check

1 in front of

2 above

Concept Check

3

 Directions: 1. Draw a ball in front of the bike. **2.** Draw a balloon above the boy.
3. Draw an X on the animal that is behind the bird.

 Directions: 4. Circle the balloon that is above, or over, the tree. **5.** Draw a box around the bird that is below, or under, the red bird. **6.** Draw an X on the butterfly that is above the flower. **7.** Draw a box around the bug that is below, or under, the other bug.

Name ....

Lesson 3
Next to and Beside

Math in My World ▶ Watch

 Teacher Directions: Circle the animal that is next to the tree. Draw an X on the animal that is beside the rock. Draw a frog next to the small elephant.

1 next to   beside

2

Directions: I. Trace the circle around the pig that is next to the orange house. Trace the X on the house that is next to the pig with the yellow shirt. **2.** Draw an X on the animal that is next to the brick house. Draw a box around the pig that is beside the pig in the red shirt. Circle the animal that is next to the blue bird. Draw a flower beside the caterpillar.

Independent Practice

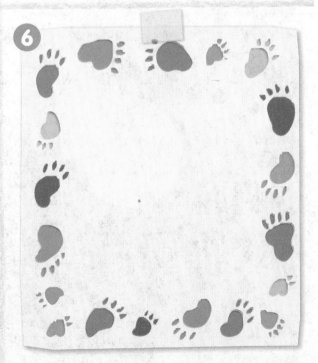

Directions: 3. Draw an X on the spoon that is next to the orange bowl. **4.** Draw an X on the chairs that are beside the green chair. **5.** Draw an X on the bed that is next to the bed with the flower blanket. **6.** Draw three different colored flowers in a row. Ask a partner to tell you which flower is next to, or beside, one of the flowers.

Directions: 7. Point to the part of the caterpillar's body that is next to the color pink. Color it orange. Point to the part of the caterpillar's body that is next to the head. Color it blue. Color all other parts any color. Tell a classmate which colors you chose, and tell the colors that are beside them.

My Homework

Homework Helper eHelp

Need help? connectED.mcgraw-hill.com

1

2

3

Directions: 1. Draw a box around the animal that is next to the fox. Draw an X on the object that is beside the rabbit. **2.** Draw an X on the animal that is beside the lizard. Draw a box around the object that is next to the turtle. **3.** Draw an X on the object that is next to the lizard. Draw a box around the animal that is next to the rabbit in the hole.

4

Vocabulary Check

5 **next to**

6 **beside**

Directions: 4. Draw an X on the animal next to the cactus. Draw a box around the animal that is beside the armadillo. **5.** Draw a bone next to the dog. **6.** Draw a ball beside the cat.

Math at Home At home, give your child directions using the words *next to* and *beside*. For example, put your shoes on the floor next to the other shoes in the closet.

Name _____

Lesson 4
Problem Solving
STRATEGY: Act It Out

Where do I put it?

Act It Out

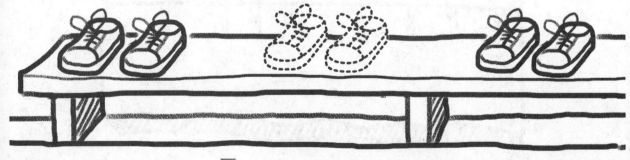

🦉 **Teacher Directions:** Use 📷 to show where the shoes should be placed. Tell where they belong. Trace the shoes to show where they belong.

Where do I put it?

Act It Out

 Directions: Use a connecting cube to show where the cat should be placed. Draw the cat where it belongs. Tell where the cat belongs.

Where do I put it?

Act It Out

Directions: Use a connecting cube to show where the lunch box should be placed. Draw the lunch box where it belongs. Tell a friend what 2 things should be above the backpacks.

Where do I put it?

Act It Out

 Directions: Use connecting cubes to show where the flowers should be placed. Draw the flowers where they belong. Tell where the flowers belong.

Lesson 4

Problem Solving:
Act It Out

My Homework

Where do I put it?

Act It Out

Directions: Use dry cereal to show where the box of cereal should be placed. Trace the box to show where it belongs. Tell where the cereal belongs.

Where do I put it?

Act It Out

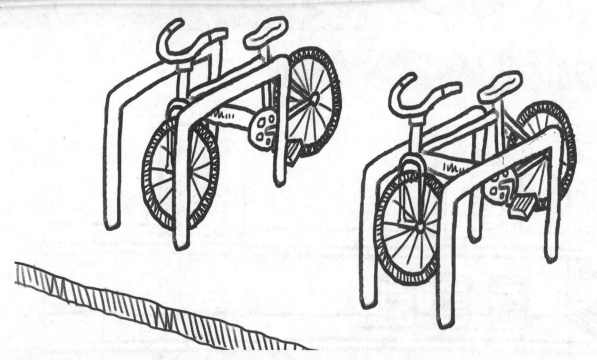

Directions: Use dry cereal to show where the bike should be placed. Draw the bike where it belongs. Tell where the bike belongs.

Math at Home Take advantage of problem-solving opportunities during daily routines such as riding in the car, bedtime, doing laundry, and so on.

Name

....

My Review

Vocabulary Check

beside

above

below

behind

next to

in front of

Directions: I. Draw another bird flying above the trees. **2.** Circle the animal behind the tree. **3.** Draw a coconut beside a monkey. **4.** Draw a monkey next to one of the lion cubs. **5.** Circle the monkey below the branch. **6.** Draw an X on the lions in front of the tree.

Concept Check

1

2

3

4

Directions: I. Draw an X on the bubble that is above, or over, the bottle. **2.** Draw a box around the flower that is below, or under, the green flower. **3.** Draw a crayon next to the box. **4.** Draw a dog bone beside the bowl.

Brain Builders

 Directions: 5. Circle the object above the dog. Draw an X on the animal that is next to the black cat. Draw a bird below the sun. Underline the object that is on the blanket next to the girl with the red book.

Reflect

 Directions: Draw an X on the butterfly that is above the flower pot. Draw a box around the animal that is in front of the baby bear. Draw a circle around the bees that are below the beehive. Circle the window that is beside the door.

Performance Task

Brain Builders

Animals Everywhere Around Us

Some animals are pets. Some animals are helpers, and some live in the forests and jungles.

Show all your work to receive full credit.

Part A

How many animals are above the bed? Circle them. Write the number.

Part B

Draw an X on the plant that is next to the deer.

Part C

How many objects are in front of the squirrel? Circle them. Write the number.

_____

Part D

Draw a box around the animal that is next to the giraffe. Draw an X on the animal that is beside the plant.

Chapter 11 Two-Dimensional Shapes

Let's Discover Shapes!

Watch a video!

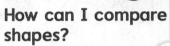

Watch ▶

Brain Builders

Chapter 11 Project

Shape Chart

1 You will make a chart to show the shapes you learned about in this chapter.

2 Help your teacher list the name of each shape you learned about on the board.

3 With a partner, make a chart showing all the shapes you learned about. Trace pattern blocks or attribute blocks to draw the shapes, or use your own drawings.

4 Write the name of each shape next to your drawing. You can use what your teacher wrote on the board to help you.

5 In the space below, draw a picture of a real-world object that matches each shape in your chart.

Name ....

Am I Ready?

1

2

3

 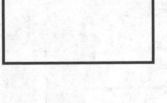 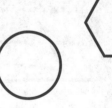

4

   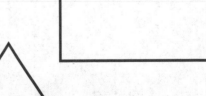

 Directions: 1–2. Circle the objects that are the same shape. **3–4.** Look at the shapes. Find the shapes in the row that are the same. Color them blue.

My Math Words

Review Vocabulary

length   size

Directions: Trace each word. Use the words to describe the objects in each box.

My Vocabulary Cards

circle

hexagon

rectangle

round

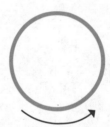

side

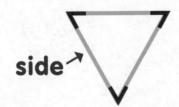

side →

straight

Teacher Directions:
Ideas for Use

- Group three words that belong together. Add a word that does not belong in the group. Ask another student to name the word that does not belong.

- Draw a line on each card every time you read or write the word in this chapter. Try to use at least 10 lines for each card.

hexagon

circle

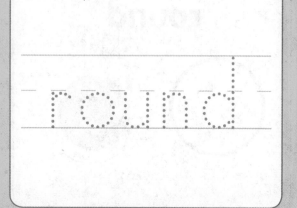

round

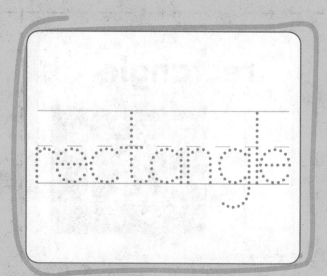

rectangle

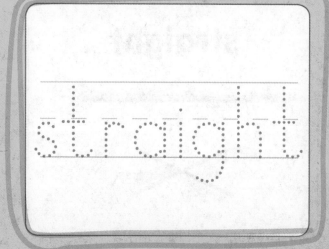

straight

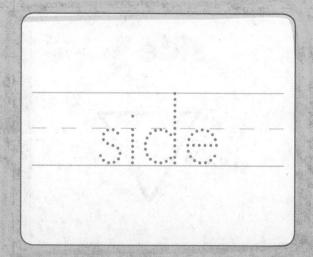

side

square

triangle

vertex

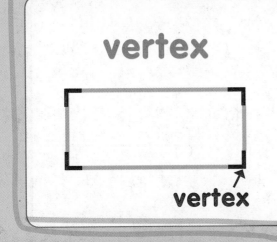

vertex

Teacher Directions:
More Ideas for Use

- Have students sort the words by the number of letters in each word.

- For each blank card, have students write a letter on the front and then draw a picture of an object that starts with that letter on the back.

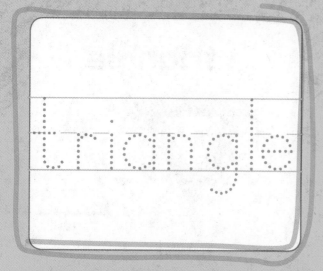

triangle

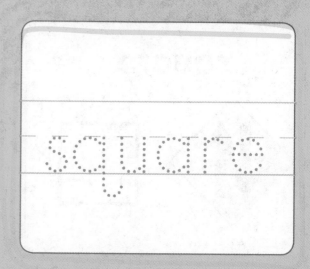

square

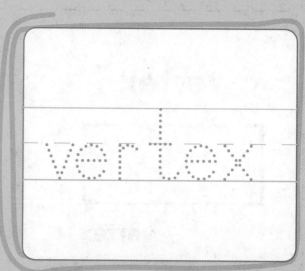

vertex

My Foldable

FOLDABLES Follow the steps on the back to make your Foldable.

square

triangle

rectangle

circle

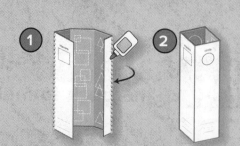

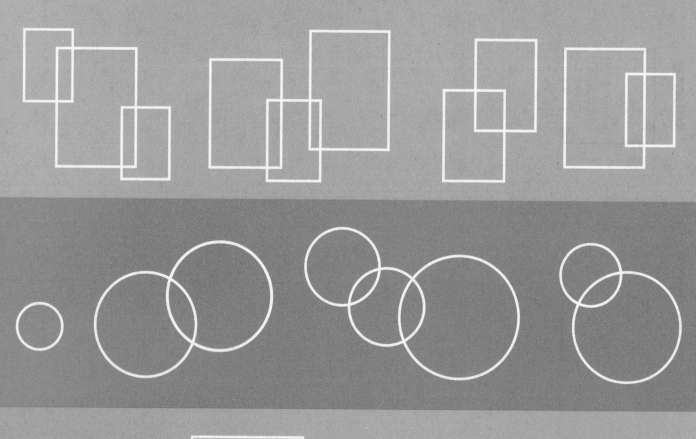

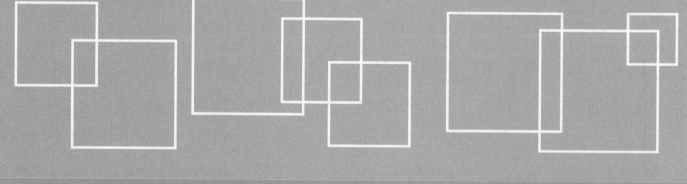

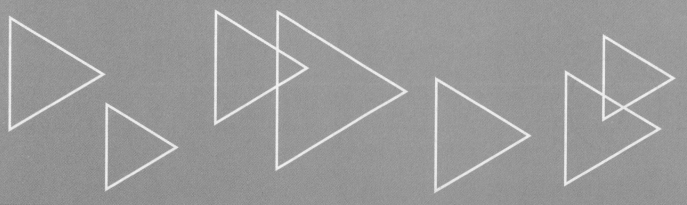

Lesson 1
Squares and Rectangles

Math in My World Tools Watch

 Teacher Directions: Use ▢ ▬ to make a picture. Trace the shapes. Identify the shapes by coloring the squares blue and rectangles orange. Describe the shapes using the words *longer sides* and *sides that are the same length*. How many sides does each shape have? How many vertices does each shape have?

Guided Practice

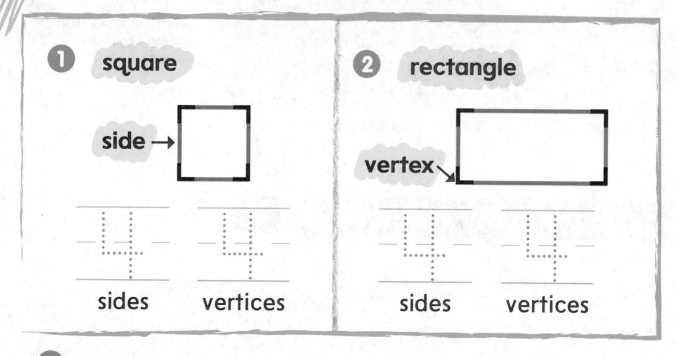

1 square

side →

____ ____
sides   vertices

2 rectangle

vertex →

____ ____
sides   vertices

3

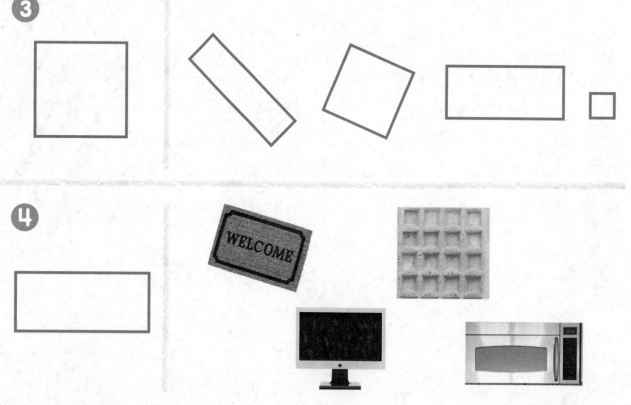

4

WELCOME

Directions: 1–2. Name the shape. Trace the number of sides and vertices. **3.** Name the shape. Describe it. Compare the shape to each shape in the group. Circle the matching shapes. **4.** Name the shape. Describe it. Compare the shape to each object in the group. Circle the objects that are the same shape.

Name

Independent Practice

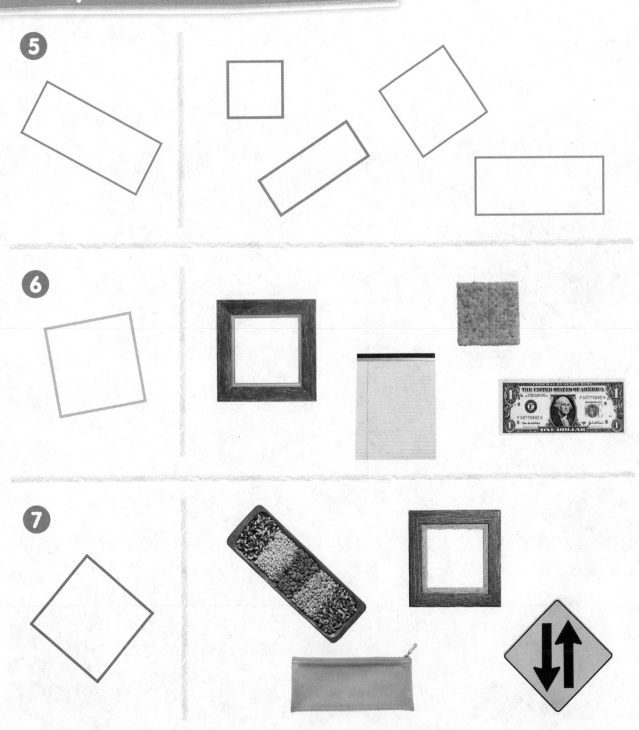

⑤

⑥

⑦

Directions: 5. Name the shape. Describe it. Compare the shape to each shape in the group. Circle the matching shapes. **6–7.** Name the shape. Describe it. Compare the shape to each object in the group. Circle the objects that are the same shape.

Brain Builders

Processes &Practices

8

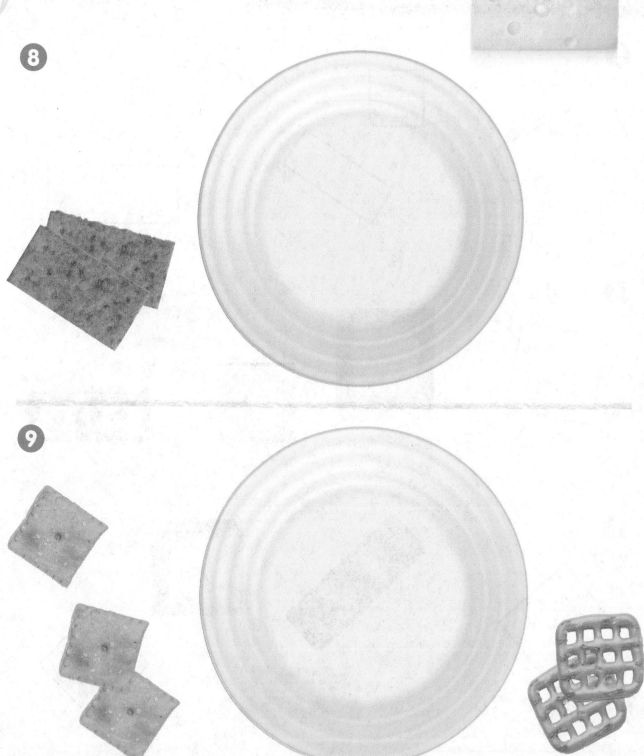

9

Directions: 8. Draw rectangle-shaped foods on the plate. **9.** Draw square-shaped foods on the plate. **10.** Explain to a friend how a square is different from a rectangle.

My Homework

Homework Helper

Need help? connectED.mcgraw-hill.com

1

 4   4   4   4

sides   vertices   sides   vertices

2

3

   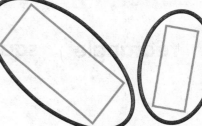 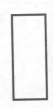

 Directions: 1. Name each shape. Write the number of sides and vertices. **2–3.** Name the shape. Describe it. Compare the shape to each shape in the group. Circle the matching shapes.

4

5

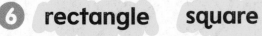

Vocabulary Check

6 rectangle   square

Directions: 4–5. Name the shape. Describe it. Compare the shape to each object in the group. Circle the objects that are the same shape. **6.** Draw a rectangle. Draw a square. Circle the shape that has sides that are all the same length.

Math at Home Look at flat objects in your home such as table tops, windows, cupboard doors, and stair steps. Ask your child to identify the shape of each.

Name

Lesson 2
Circles and Triangles

Math in My World   Tools   Watch

 Teacher Directions: Use ● ▲ to make a picture. Trace the shapes. Identify the shapes by coloring the circles red and the triangles purple. Describe the shapes using the words *round* and *straight*. Count how many sides and vertices each shape has.

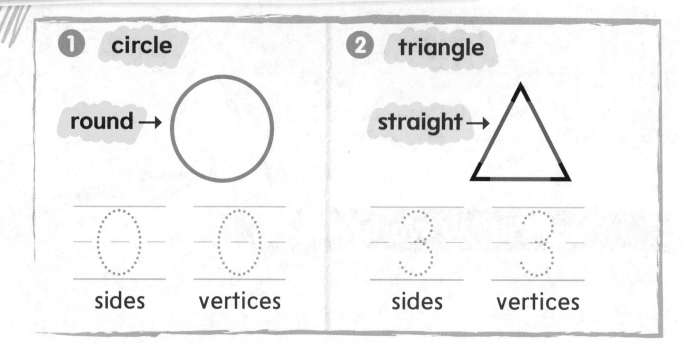

1 circle

round →

sides   vertices

2 triangle

straight →

sides   vertices

3

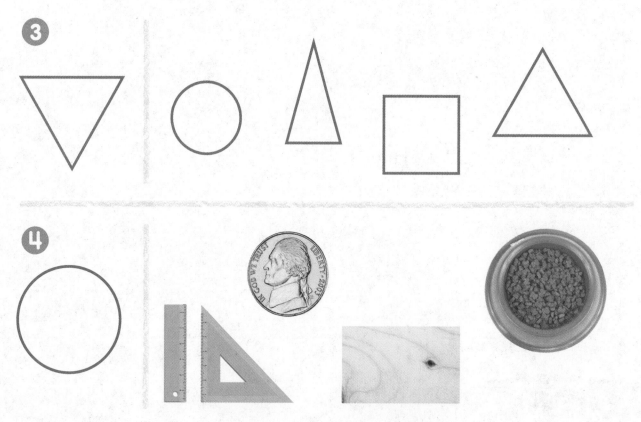

4

Directions: 1–2. Name the shape. Trace the number of sides and vertices. **3.** Name the shape. Describe it. Compare the shape to each shape in the group. Circle the matching shapes. **4.** Name the shape. Describe it. Compare the shape to each object in the group. Circle the objects that are the same shape.

Independent Practice

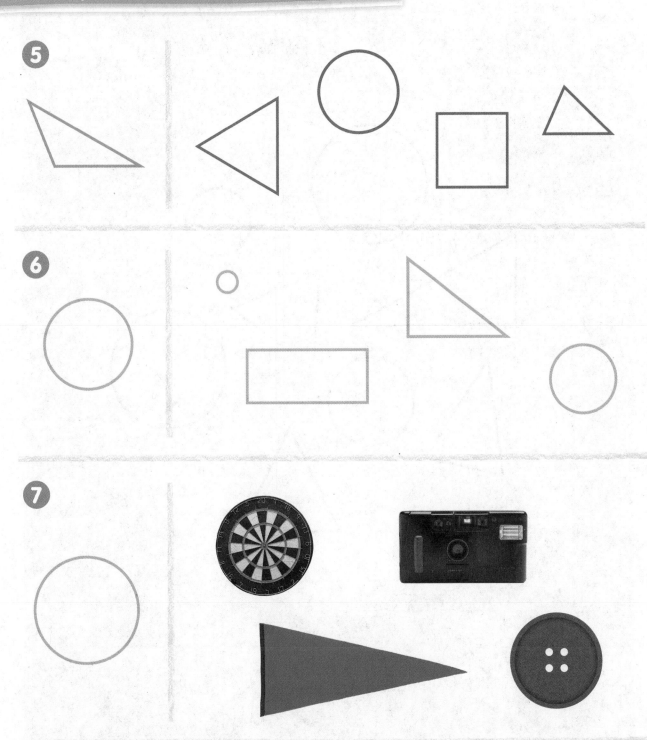

5

6

7

Directions: 5–6. Name the shape. Describe it. Compare the shape to each shape in the group. Circle the matching shapes. **7.** Name the shape. Describe it. Compare the shape to each object in the group. Circle the objects that are the same shape.

 Directions: 8. Identify the circles and triangles on the fruit pizza by coloring the circles green and the triangles orange. Explain to a friend how you can tell the difference between a triangle and a circle.

My Homework

Homework Helper

Need help? connectED.mcgraw-hill.com

1

0	**0**	**3**	**3**
sides	vertices	sides	vertices

2

3

 Directions: I. Name each shape. Write the number of sides and vertices.
2–3. Name the shape. Describe it. Compare the shape to each shape in the group.
Circle the matching shapes.

4

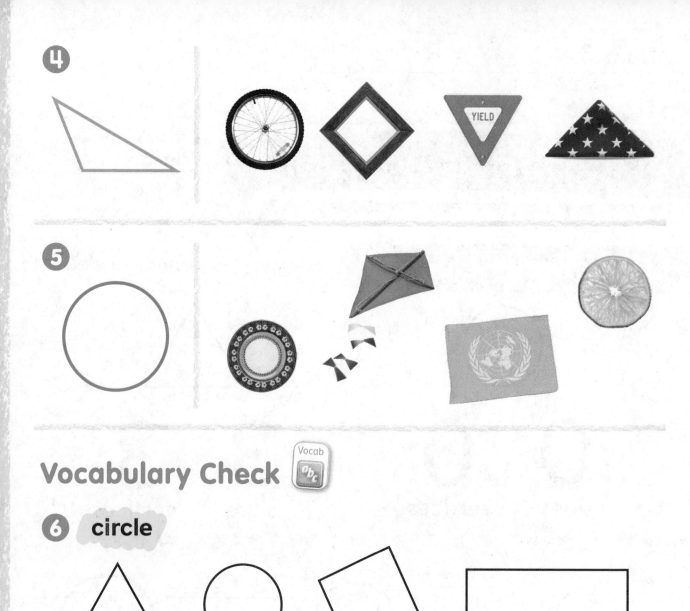

5

Vocabulary Check

6 circle

7 triangle

Directions: 4–5. Name the shape. Describe it. Compare the shape to each object in the group. Circle the objects that are the same shape. **6.** Color the circle red. **7.** Color the triangle blue.

Math at Home Gather a group of objects of various shapes and sizes. Include triangles and circles. Ask your child to sort the triangles and circles.

Name

ESSENTIAL QUESTION
How can I compare shapes?

 Math in My World Tools Watch

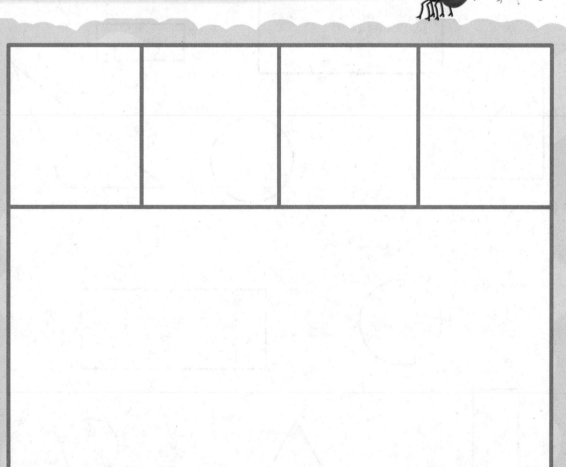

 Teacher Directions: Place  at the bottom of the mat. Sort the attribute blocks. Place one of each shape in the boxes at the top. Trace the shapes. Name the shapes. Count the sides and vertices of each shape. Compare the shapes.

1

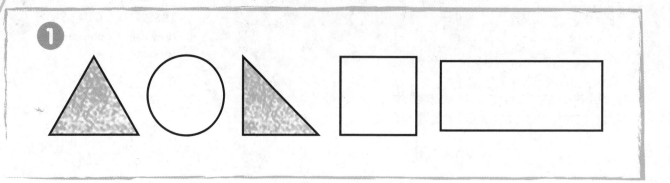

2

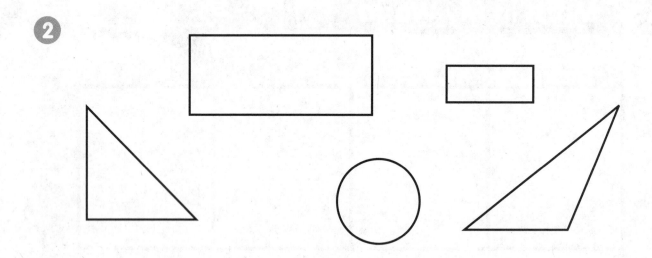

3

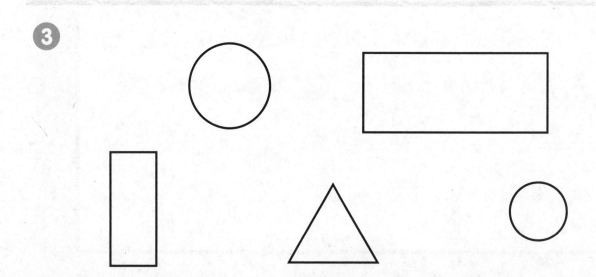

Directions: I. Color the shapes that have three sides and three vertices blue.
2. Color the shapes that have four sides and four vertices red. **3.** Color the shapes that have zero sides and zero vertices green.

Independent Practice

4

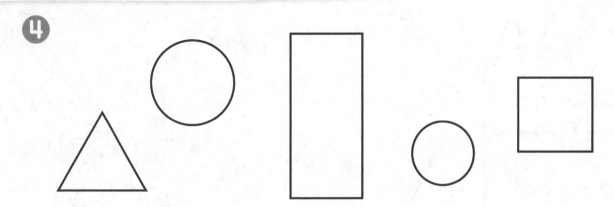

5

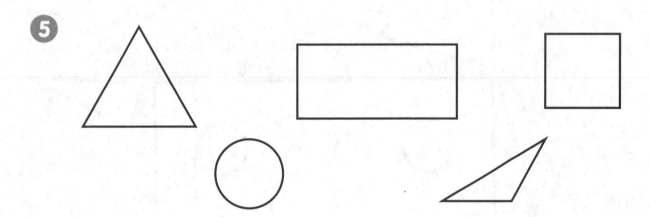

6

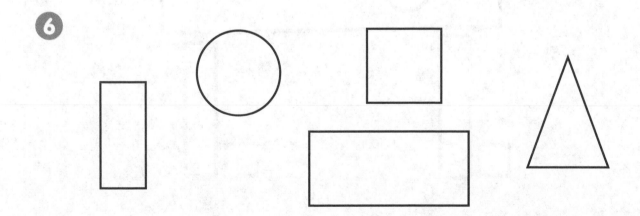

Directions: 4. Color the shapes that are round purple.
5. Color the shapes that have three sides and three vertices red.
6. Color the shapes that have four sides and four vertices orange.

7

 Directions: 7. Color the square(s) red, the triangle(s) yellow, the rectangle(s) green, and the circle(s) blue. Draw Xs on the shapes with less than 3 vertices.

My Homework

Homework Helper eHelp

Need help? connectED.mcgraw-hill.com

1

square   circle   triangle   rectangle

2

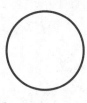 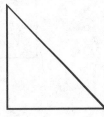 

3

   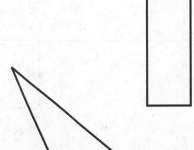

 Directions: 1. Color the shapes that have four sides and four vertices blue.
2. Color the shapes that have three sides and three vertices red.
3. Color the shapes that have zero sides and zero vertices green.

4

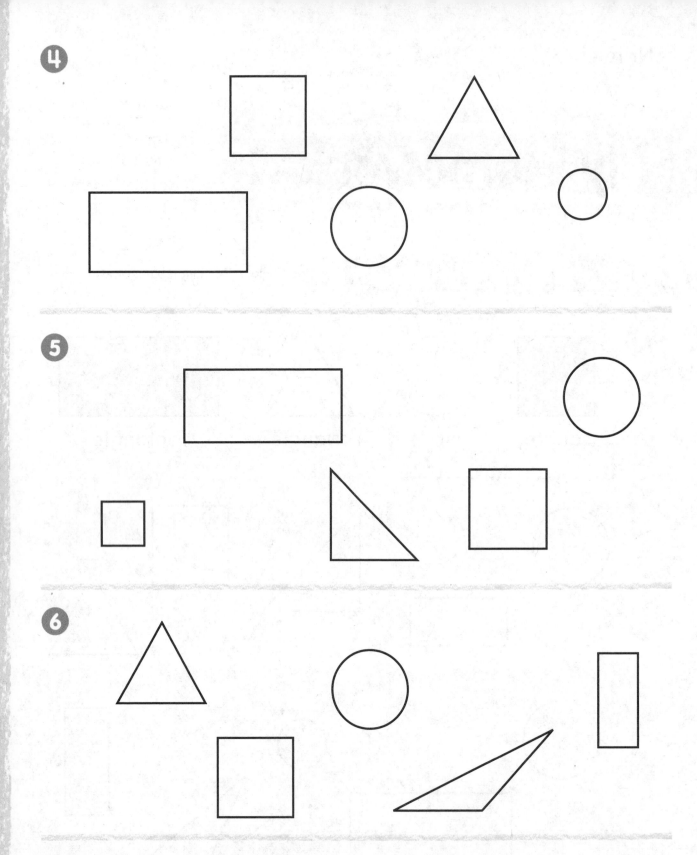

5

6

Directions: 4. Color the shapes that are round purple.
5. Color the shapes that have four sides and four vertices yellow.
6. Color the shapes that have three sides and three vertices orange.

Math at Home While driving in the car, have your child look for shapes. For example, street signs, wheels, etc. Have him or her identify each shape seen.

Name

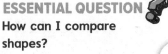

Lesson 4
Hexagons

ESSENTIAL QUESTION
How can I compare shapes?

 Math in My World Tools Watch

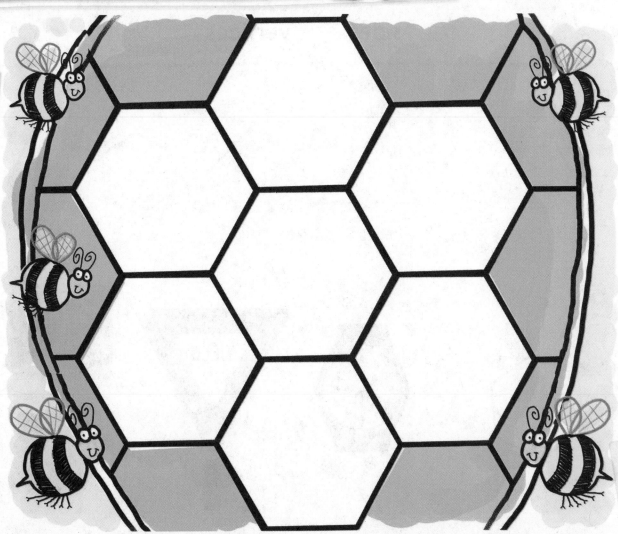

 Teacher Directions: Use ⬡ to fill in the beehive. Color each hexagon a different color. Count the sides and vertices. Describe the shape using the words *sides* and *vertices*.

1 hexagon

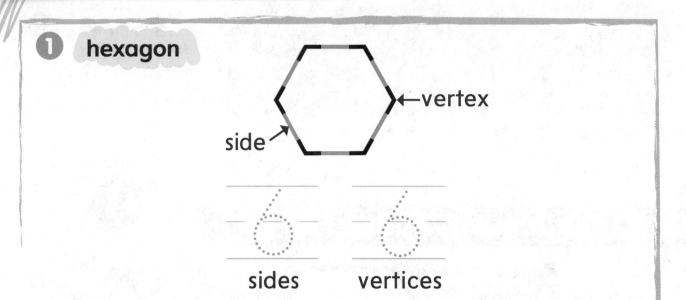

← vertex

side →

6   6

sides   vertices

2

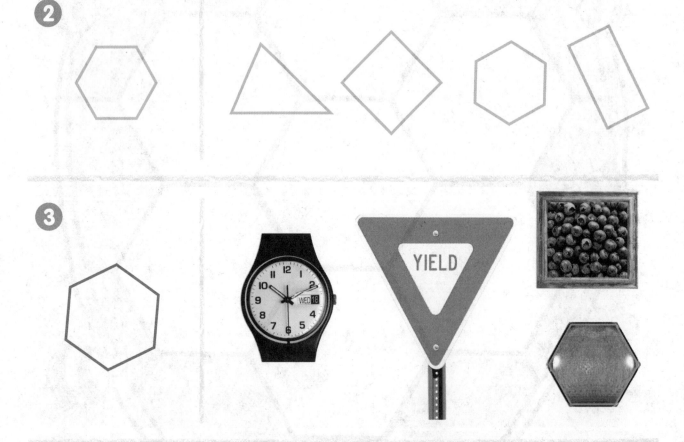

3

YIELD

Directions: 1. Name the shape. Trace the number of sides and vertices. **2.** Name the shape. Describe it. Compare the shape to each shape in the group. Circle the matching shape. **3.** Name the shape. Describe it. Compare the shape to each object in the group. Circle the object that is the same shape.

Name

Independent Practice

 4

 5

 6

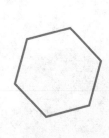   

 Directions: 4. Name the shape. Describe it. Compare the shape to each shape in the group. Circle the matching shape. **5–6.** Name the shape. Describe it. Compare the shape to each object in the group. Circle the object that is the same shape.

7

 Directions: 7. Use hexagon and triangle pattern blocks to create flowers. Trace the blocks. Color the hexagons orange. Color the triangles blue. Name the shapes you used. Write the number of sides and vertices on each shape.

My Homework

Homework Helper

Need help? connectED.mcgraw-hill.com

1

←vertex

side→

_____ _____

_____ **6**

sides   vertices

2

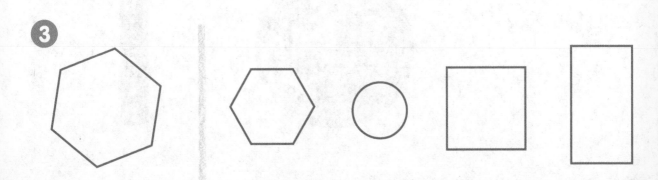

3

 Directions: 1. Name the shape. Write the number of sides and vertices. **2–3.** Name the shape. Describe it. Compare the shape to each shape in the group. Circle the matching shape.

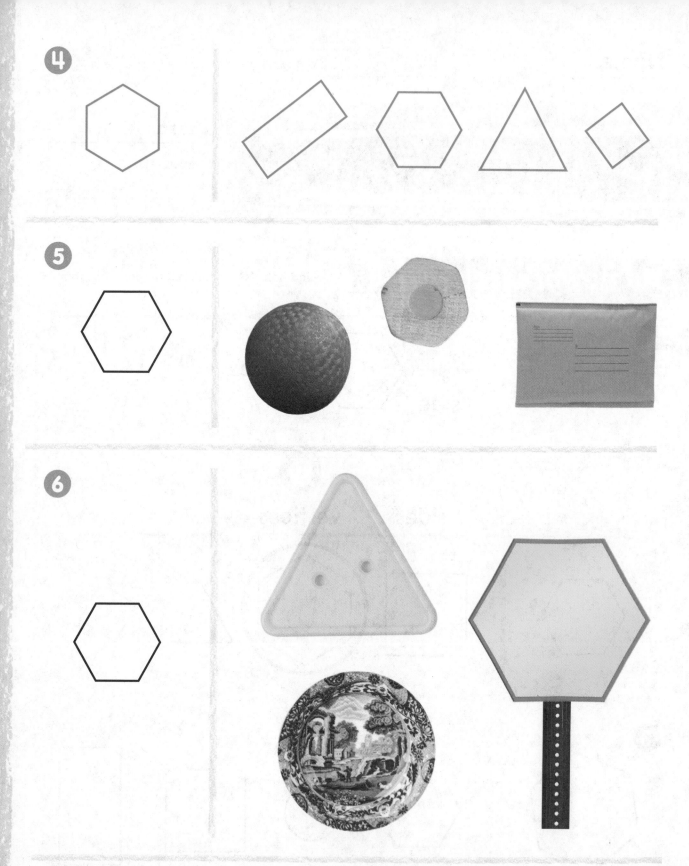

4 **5** **6**

Directions: 4. Name the shape. Describe it. Compare the shape to each shape in the group. Circle the matching shape. **5–6.** Name the shape. Describe it. Compare the shape to each object in the group. Circle the object that is the same shape.

Math at Home While in the grocery store, have your child look to see if he or she can see shapes. Have your child look for hexagons.

Name _____

Check My Progress

Vocabulary Check

1 circle

2 triangle

Concept Check

3
  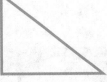

4
 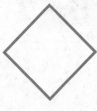   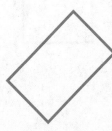

 Directions: I. Draw circles and color them blue. **2.** Draw triangles and color them orange. **3–4.** Name the shape. Describe it. Compare the shape to each shape in the group. Circle the matching shapes.

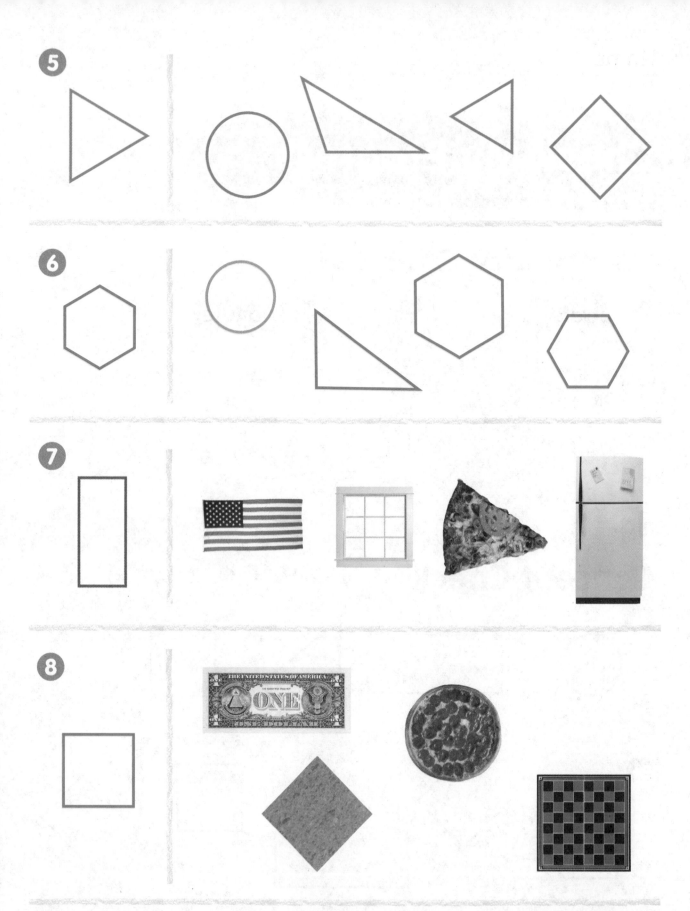

5

6

7

8

 Directions: 5–6. Name the shape. Describe it. Compare the shape to each shape in the group. Circle the matching shapes. **7–8.** Name the shape. Describe it. Compare the shape to each object in the group. Circle the objects that are the same shape.

Name ....

Lesson 5
Shapes and Patterns

Math in My World   Tools   Watch

Teacher Directions: Use pattern blocks to copy the pattern. Color the shapes to match the pattern. Circle the shape that could come next.

1

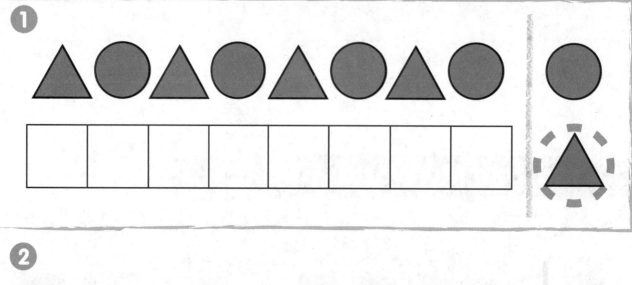

2

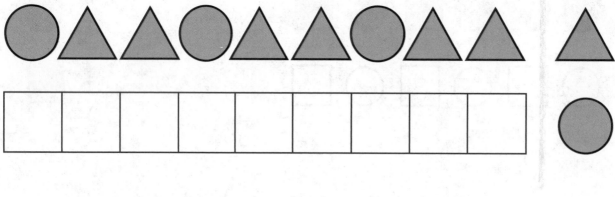

3

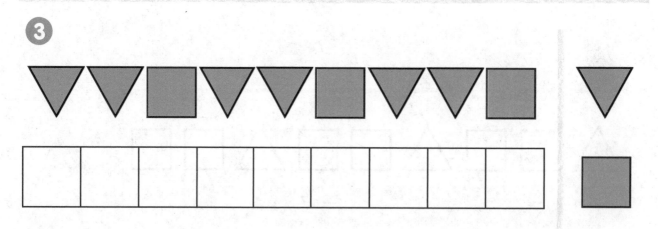

Directions: 1. Identify the pattern. Draw the shapes to copy the pattern. Draw a circle around the shape that could come next. Tell how you know. **2–3.** Identify the pattern. Draw the shapes to copy the pattern. Draw a circle around the shape that could come next.

Name _____

4

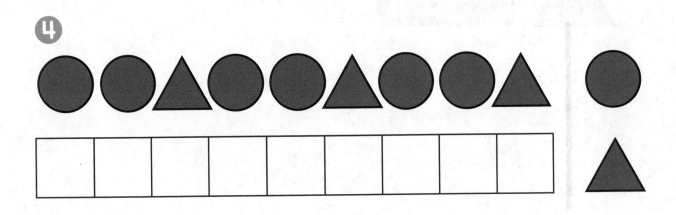

5

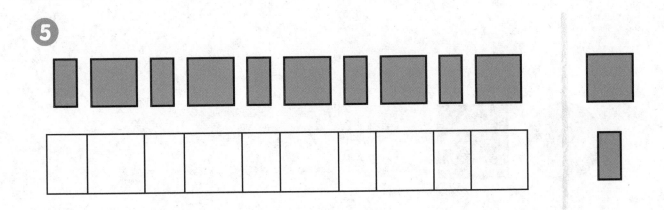

6

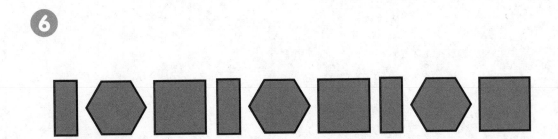

 Directions: 4–5. Identify the pattern. Draw the shapes to copy the pattern. Draw a circle around the shape that could come next. Tell how you know. **6.** Identify the pattern. Draw the shape that could come next. Tell a friend how you know.

7

8

 Directions: 7–8. Use the shapes given to create a pattern. Draw and color your pattern. Describe your pattern to a friend.

Name

My Homework

Homework Helper

Need help? connectED.mcgraw-hill.com

1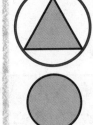

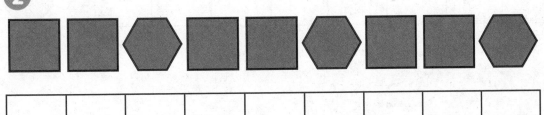 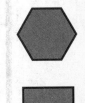

2

3

 Directions: 1–3. Identify the pattern. Draw the shapes to copy the pattern. Circle the shape that could come next.

4

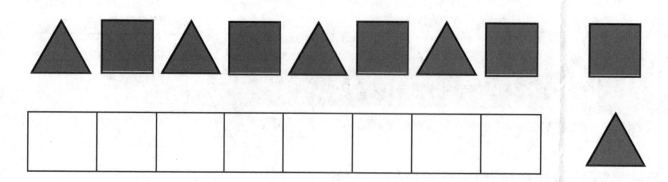

5

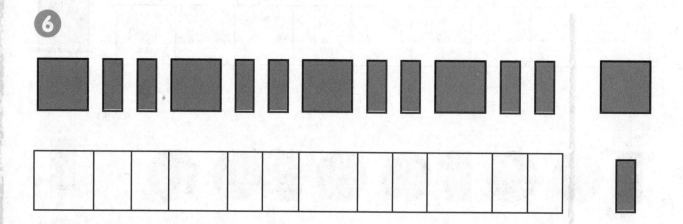

6

Directions: 4–6. Identify the pattern. Draw the shapes to copy the pattern. Circle the shape that could come next.

Math at Home Use shapes to make a pattern. Have your child copy the pattern and tell what could come next in the pattern.

Name ....

 ## Math in My World

 Teacher Directions: Use ▮ ▮ ◯ ⬡ . Place an attribute block on each object that has a matching shape. Describe it. Circle the object that is *above* the desk. Draw a square picture *above* the bed. Draw a rectangle basket *in front of* the bed.

1

2

Directions: 1. Circle the object that is *beside* the rectangle picture. Describe its shape. Draw an X on the object that is above the box. Describe its shape. **2.** Circle the object that is *above* the barn. Describe its shape. Draw an X on the object that is *in front* of the barn. Describe its shape.

Independent Practice

3

4

 Directions: 3. Circle the object that is *below* the table. Describe its shape. Draw an X on the object that is *next to* the box. Describe its shape. Draw a box around the object that is *above* the box. Describe its shape. **4.** Circle the object that is *above* the clipboard. Describe its shape. Draw a box around the object that is *next to* the racket. Describe its shape.

5

6

Directions: 5. Draw a window *above* the door and *next to* the other window.
Describe its shape. **6.** Draw a swing *behind* the dog and *next to* the other swing.
Describe its shape.

Name

My Homework

Homework Helper eHelp

Need help? connectED.mcgraw-hill.com

1

2

Directions: 1. Circle the object that is *above* the sink. Describe its shape. Draw a box around the object that is *next to* the tissue box. Describe its shape. **2.** Circle the object that is *beside* the keyboard. Describe its shape. Draw an X on the object that is *below* the table. Describe its shape.

3

4

Directions: 3. Circle the object that is *next to* the refrigerator and *above* the cupboard. Describe its shape. Draw a box around the object that is *above* the sink. Describe its shape. Draw an X on the object that is *below* the sink. Describe its shape. **4.** Circle the object that is *next to* the basketball hoop. Describe its shape. Draw an X on the object that is *next to* the cone in the circle. Describe its shape to a family member.

Math at Home Find small circle, triangle, rectangle, hexagon, and square objects. Arrange them and have your child describe their positions.

Name

Lesson 7
Compose New Shapes

ESSENTIAL QUESTION
How can I compare shapes?

 Math in My World Tools Watch

 Teacher Directions: Use ▮ to make a large rectangle. Then make a large square. Trace one of the shapes you made.

1

2

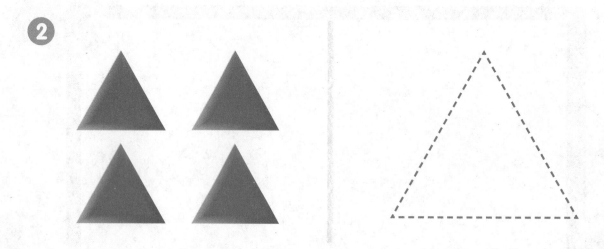

3

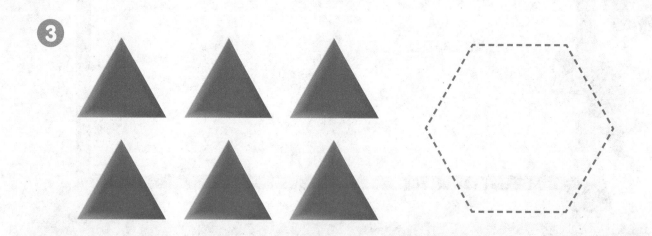

Directions: 1–3. Use the pattern blocks shown to make the larger shape. Trace the larger shape. Name the larger shape.

Independent Practice

4

5

6

 Directions: 4–5. Use the pattern blocks shown to make the larger shape. Trace the larger shape. Name the larger shape. **6.** Use nine pattern blocks to make a larger square. Trace the shapes you used.

Brain Builders

7

8

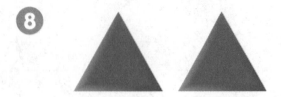

Directions: 7. Look at the rectangle. Draw the shapes that make up the rectangle. **8.** Look at the triangles. Can these two triangles make a rectangle? Try it using pattern blocks. Circle the triangles if yes. Put an X on the triangles if no.

Name ....

My Homework

Homework Helper

Need help? connectED.mcgraw-hill.com

1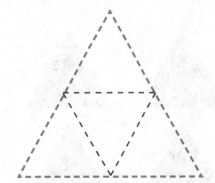

2

3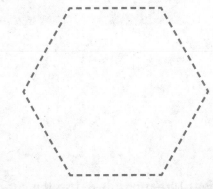

 Directions: 1–3. Using *Manipulatives Masters: Pattern Blocks, Sheet I,* color the pattern block triangles green and the squares orange. *Note: Parents, cut out the paper shapes for your child.* Use the cut-out shapes to make the larger shape. Name the shape. Color it.

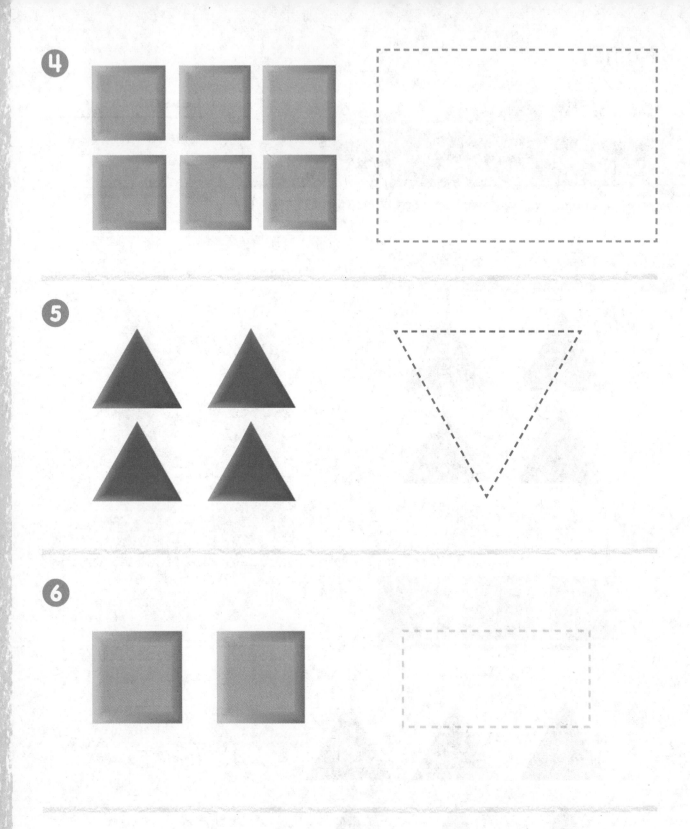

4

5

6

Directions: 4–6. Use the cut-out shapes to make the the larger shape. Name the shape. Color it.

Math at Home Have your child count 8 cut-out squares. Direct your child to make a rectangle with the squares.

Name ....

Lesson 8
Problem Solving
STRATEGY: Use Logical Reasoning

ESSENTIAL QUESTION
ESSENTIAL QUESTION
How can I compare shapes?

What shapes are missing?

Use Logical Reasoning

Directions: Use pattern blocks to find the missing shapes. Name and describe the shapes in the larger shape. Trace the shapes.

Use Logical Reasoning

 Directions: Use pattern blocks to find the missing shapes. Name and describe the shapes in the larger shape. Draw the shapes.

What shapes are missing?

Use Logical Reasoning

 Directions: Use pattern blocks to find the missing shapes. Name and describe the shapes in the larger shape. Draw the shapes. Explain to a friend how you found the missing shapes.

What shapes are missing?

Use Logical Reasoning

 Directions: Use pattern blocks to find the missing shapes. Name and describe the shapes in the larger shape. Draw the shapes.

My Homework

What shapes are missing?

Use Logical Reasoning

 Directions: Use patterns blocks from *Manipulative Masters: Pattern Blocks, Sheet I* to find the missing shapes. Name and describe the missing shapes. Trace the missing shapes. *Note: Parents, cut out the paper shapes for your child.*

What shapes are missing?

Use Logical Reasoning

Directions: Use pattern block cut outs to find the missing shapes. Name and describe the missing shapes. Draw the missing shapes.

Math at Home Take advantage of problem-solving opportunities during daily routines such as going to the park. Have your child identify shapes at the park.

Name ....

ESSENTIAL QUESTION
How can I compare shapes?

Math in My World

Tools Watch

Teacher Directions: Use to make a new shape. Trace the shapes to show your new shape. Describe the new shape to a partner.

1

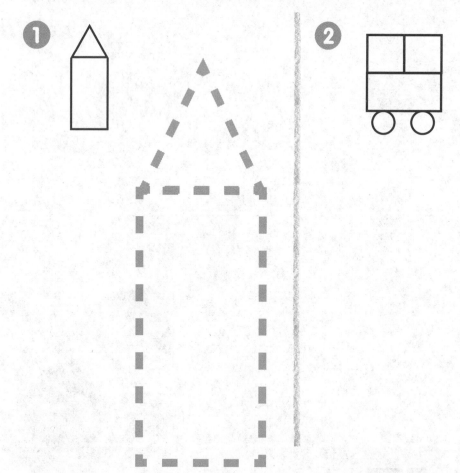

2

Directions: 1–2. Look at the picture. What shapes make the picture? Name the shapes. Use attribute blocks to model the object in the picture. Trace the attribute blocks.

Name

_____

③

④

Directions: 3-4. Look at the picture. What shapes make the picture? Name the shapes. Use attribute blocks and pattern blocks to model the object in the picture. Trace the attribute blocks and pattern blocks.

5

Directions: 5. Use square, rectangle, circle, triangle, and hexagon pattern blocks and attribute blocks to make fish in the bowl. Trace the shapes and color them. Then use the blocks to make an object *in front of* the bowl. Tell a partner the names of the shapes you used.

Name _____

Homework Helper eHelp 🏠

Need help? connectED.mcgraw-hill.com

1

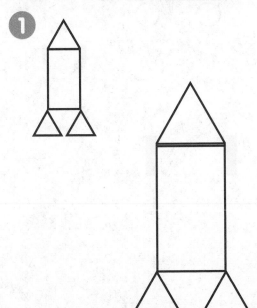

2

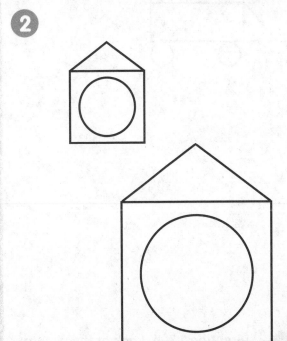

Directions: 1–2. Look at the object. What shapes make the object? Name the shapes. Draw the shapes used to model the object in the picture. Color the shapes.

3

4

Directions: 3–4. Look at the picture. What shapes make the picture? Name the shapes. Draw the shapes used to model the object in the picture. Color the shapes.

Math at Home Choose another object in the picture above. Have your child draw the shapes in the picture above to model the object.

Name _____

My Review

Vocabulary Check

1 rectangle

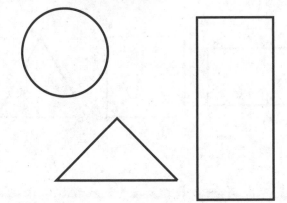

2 hexagon

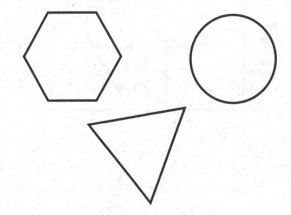

3 square

4 round

Directions: 1. Color the rectangle red. **2.** Color the hexagon blue. **3.** Color the square orange. **4.** Color the shape that is round purple.

Concept Check

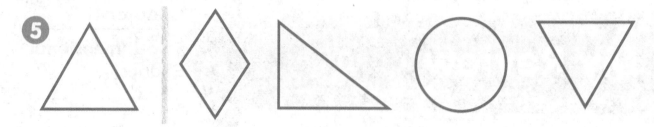

5

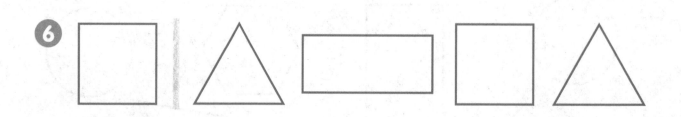

6

7

Directions: 5–6. Name the shape. Describe it. Compare the shape to each shape in the group. Circle the matching shapes. **7.** Look at the picture. What shapes do you see? Use attribute blocks to model the car. Trace the shapes.

Brain Builders

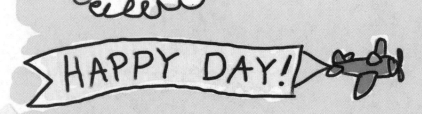

 Directions: 8. Look at the picture. Circle the object that is *above* the tree. Describe its shape. Draw an X on the object that is *next to* the tree. Describe its shape. Draw a wagon *in front of* the girls. Describe its shape.

sides   vertices

sides   vertices

sides   vertices

sides   vertices

Directions: Trace each shape. Identify and describe the shapes. Write how many sides and vertices.

Performance Task

Brain Builders

Shapes Art

Abby uses shapes to make her artwork.

Part A

Different shapes are shown. Circle the triangles. Underline the hexagon. Put an X on the squares.

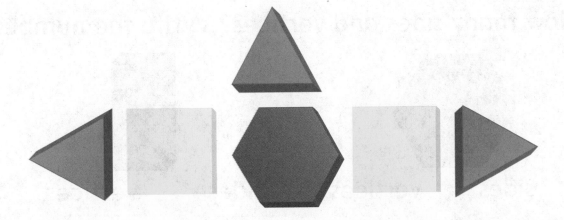

Part B

Abby makes a pattern with her shapes. Draw the two shapes that come next.

Part C

Abby made a house using shapes. How many of each shape did she use? Write the numbers.

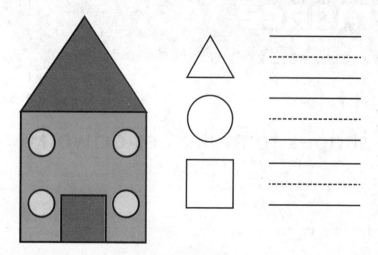

_____

- - - - - - - - -

_____

_____

_____

- - - - - - - - -

_____

Part D

How many sides and vertices? Write the numbers.

sides   vertices   sides   vertices

_____   _____   _____   _____

- - - -   - - - -   - - - -   - - - -

_____   _____   _____   _____

Chapter 12

Three-Dimensional Shapes

ESSENTIAL QUESTION

How do I identify and compare three-dimensional shapes?

My Dreams Take Me Places!

Watch a video!

Watch

Name _____

Chapter 12 Project

Shape Museum Display

1 Bring empty containers and objects from home that represent various three-dimensional shapes you learned about in the chapter.

2 Display these items along with other students' items in the Shape Museum your teacher has set up for you.

3 During this chapter, each day your teacher will give you time to visit the Shape Museum and sort the containers and objects depending on their size, shape, and other attributes.

4 Draw pictures of some of the items in the Shape Museum and label each item with a vocabulary word that describes it like *round, cube, sphere, cone, cylinder, circle, triangle, rectangle, roll, slide,* or *stack.*

Name
_____

Am I Ready?

1

2

3

4

 Directions: 1–4. Circle the shapes in the row that are the same.

My Math Words

Review Vocabulary

circle   square

 Directions: Trace each word and tell its meaning. Look at the shapes that make the objects in the clouds. Color the squares red. Draw an X on each circle. Tell how you know.

My Vocabulary Cards

Vocab

Processes & Practices

cone

cube

cylinder

roll

slide

sphere

Teacher Directions:
Ideas for Use

- Direct students to create riddles for each word. Ask them to work with a classmate to guess the words for each riddle.

- Guide students to read each shape word. Have students describe each shape and tell how each is alike and different.

cube

cone

roll

cylinder

sphere

slide

stack

Teacher Directions:
More ideas for Use

- Direct a student to act out *roll, stack,* or *slide* with a partner. Have another student find the vocabulary word that indicates the action acted out.

- Instruct students to use the blank cards to create alphabet cards. Suggest that students write a letter on the front of a card and then write a math word that begins with that letter on the back.

stack

My Foldable

FOLDABLES® Follow the steps on the back to make your Foldable.

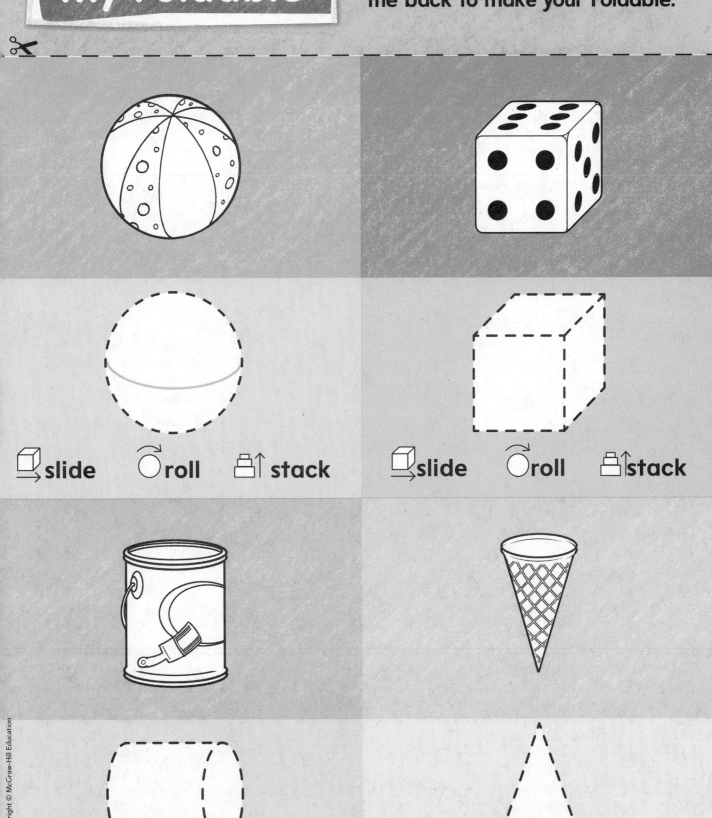

slide   roll   stack

slide   roll   stack

slide   roll   stack

slide   roll   stack

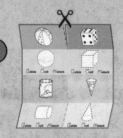
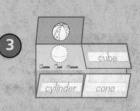

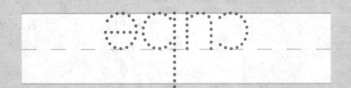

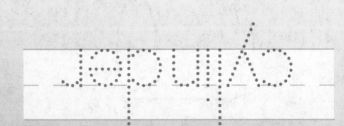

Name
....

ESSENTIAL QUESTION
How do I identify and compare three-dimensional shapes?

 Math in My World ▶ Watch

 Teacher Directions: Go on a three-dimensional shape walk around the classroom. Identify objects that are shaped like spheres and cubes. Draw a picture of one of the objects beside the matching shape. Tell whether the objects are solid or flat.

1 sphere

2 cube

3

4

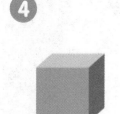

Directions: 1–2. Name the shape above the objects. Describe it. Compare it to the shapes of the objects below it. Trace the circle around the matching shape. **3–4.** Name the first shape in the row. Describe it. Compare it to the shapes of the objects in the row. Circle the matching shape.

Name

Independent Practice

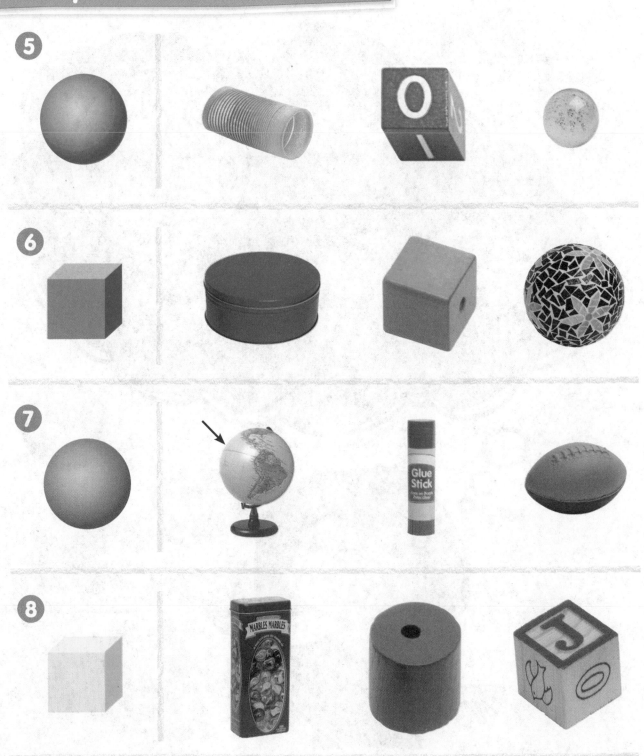

5

6

7

Glue
Stick

8

MARBLES MARBLES

Directions: 5–8. Name the first shape in the row. Describe it. Compare it to the shapes of the objects in the row. Circle the matching shape.

Brain Builders

9

Directions: 9. Name the shape of the planet. Find those shapes on the page. Draw lines from those shapes to the planet. Identify the part of the rocket that is shaped like a cube. Find those shapes on the page. Draw lines from those shapes to the rocket. Explain to a friend what a cube looks like.

My Homework

Homework Helper

eHelp

Need help? connectED.mcgraw-hill.com

1

2

3

4

Directions: 1–4. Name the first shape in the row. Describe it. Compare it to the shapes of the objects in the row. Circle the matching shape.

6

Vocabulary Check

7 sphere

8 cube

Directions: 5–6. Name the first shape in the row. Describe it. Compare it to the shapes of the objects in the row. Circle the matching shape. **7.** Draw an X on the objects that are not shaped like a sphere. **8.** Draw an X on the objects that are not shaped like a cube.

Math at Home Discuss the spheres and cubes shown on the My Homework pages. Help your child create a list of objects shaped like spheres and cubes and draw the objects.

Name ....

Lesson 2
Cylinders and Cones

ESSENTIAL QUESTION
How do I identify and compare three-dimensional shapes?

 **Math in My World** ▶ Watch

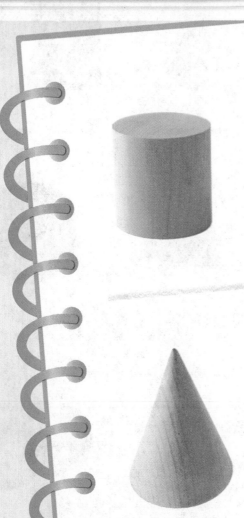

Teacher Directions: Go on a three-dimensional shape walk around the classroom. Identify objects that are shaped like cylinders and cones. Draw a picture of one of the objects beside the matching shape. Tell whether the objects are solid or flat.

1 cylinder

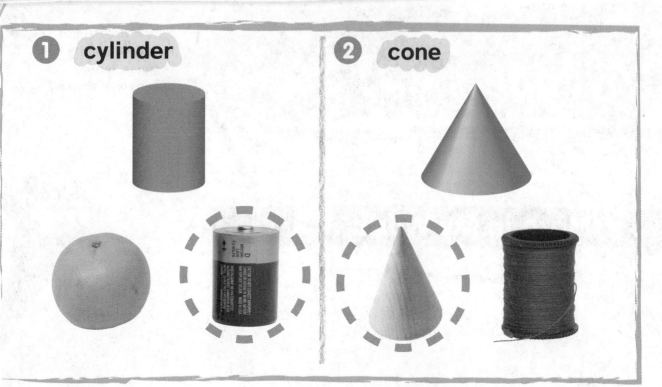

2 cone

3

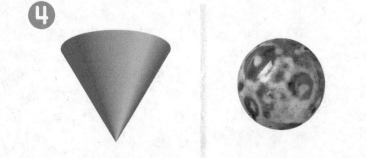

4

Directions: 1–2. Name the shape above the objects. Describe it. Compare it to the shapes of the objects below it. Trace the circle around the matching shape. **3–4.** Name the first shape in the row. Describe it. Compare it to the shapes of the objects in the row. Circle the matching shape.

Name

Independent Practice

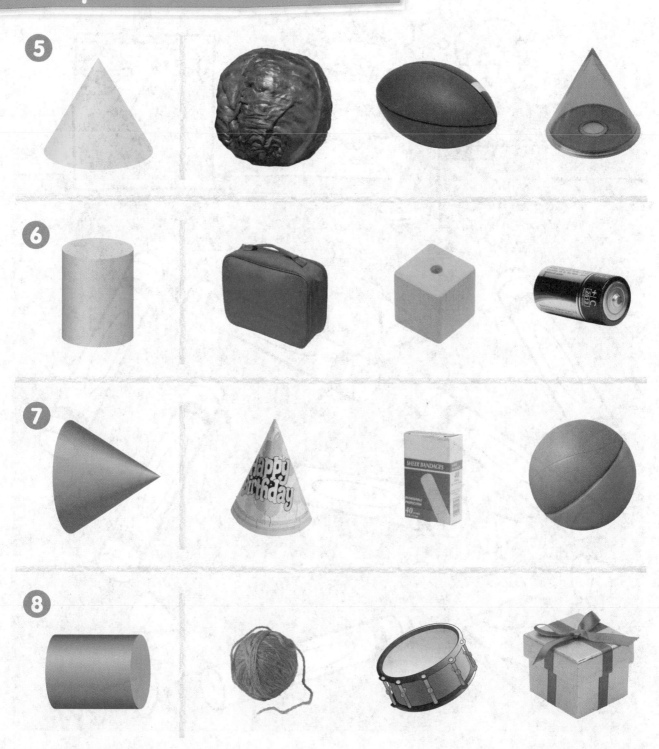

5

6

7

8

Directions: 5–8. Name the first shape in the row. Describe it. Compare it to the shapes of the objects in the row. Circle the matching shape.

Processes & Practices

Directions: 9. Point to the broken pieces of crayons. Name the shape of each piece. Identify each shape by coloring the cones blue and the cylinders red. Explain to a friend what a cylinder looks like.

Name

My Homework

Homework Helper [eHelp]

Need help? connectED.mcgraw-hill.com

1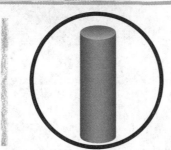

2

3

4

Directions: 1–4. Name the first shape in the row. Describe it. Compare it to the shapes of the objects in the row. Circle the matching shape.

5

6

Vocabulary Check

7 cylinder

8 cone

Copyright © McGraw-Hill Education

 Directions: 5–6. Name the first shape in the row. Describe it. Compare it to the shapes of the objects in the row. Circle the matching shape. **7.** Draw an object that is shaped like a cylinder. Tell what you drew. **8.** Draw an object that is shaped like a cone. Tell what you drew.

Math at Home Have your child find objects in your home shaped like cylinders and cones.

Name

Lesson 3
Compare Solid Shapes

ESSENTIAL QUESTION
How do I identify and compare three-dimensional shapes?

 Math in My World Tools Watch

 Teacher Directions: Use , ▪, ▮, and 🌢 to identify and compare solid shapes. Show which shapes roll, slide, and can have another shape stacked on it. Name an object in the classroom that rolls, slides, or stacks the same as one of the solid shapes. Draw the object. Tell how you know.

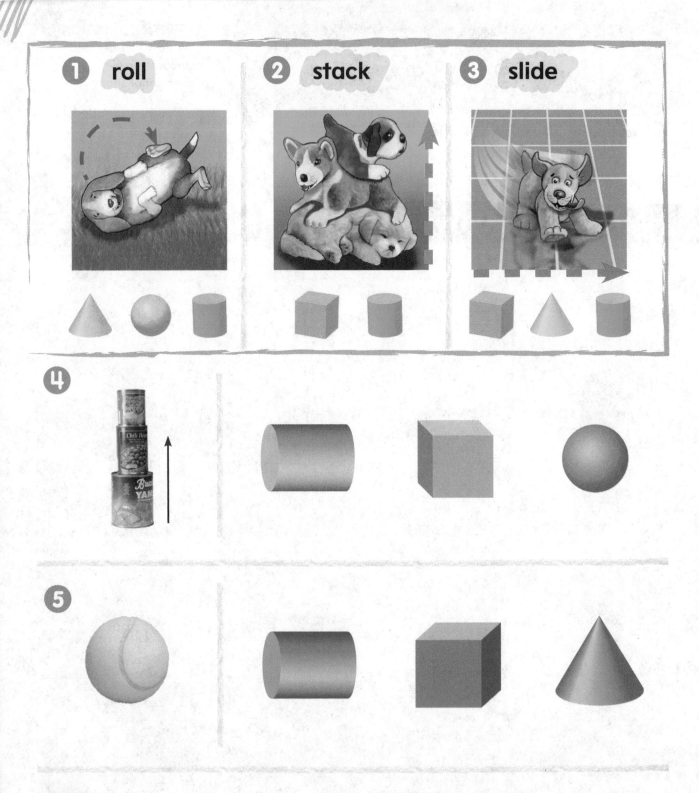

① roll

② stack

③ slide

④

⑤

 Directions: 1–3. Trace the arrows to show roll, stack, and slide. Name the shapes shown that roll, stack, and slide. **4.** Identify the shape of the first objects in the row. Describe it. Circle the other shape(s) in the row that stack. **5.** Identify the shape of the first object in the row. Describe it. Circle the other shape(s) in the row that stack.

Independent Practice

6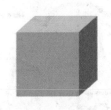

7

8

9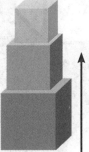

 Directions: 6. Identify the shape of the first object in the row. Describe it. Circle the other shape(s) in the row that slide. **7.** Identify the shape of the first object in the row. Describe it. Circle the other shape(s) in the row that roll. **8.** Draw an object that would roll as a cylinder rolls. **9.** Draw an object that would stack as the cubes are stacked.

Online Content at connectED.mcgraw-hill.com   Chapter 12 • Lesson 3

Directions: 10. Look at the shapes on the carpet. Find a shape that rolls, stacks, and slides. Color it orange. Find a shape that only slides and rolls and cannot have a shape stacked on it. Color it purple. Find a shape that only slides and stacks. Color it yellow. Find a shape that only rolls. Color it blue.

My Homework

Homework Helper

Need help? connectED.mcgraw-hill.com

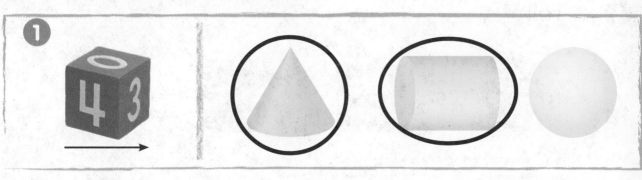

1

2

3

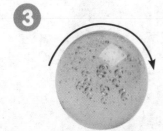

 Directions: 1. Identify the shape of the first object in the row. Describe it. Circle the other shape(s) in the row that slide. **2.** Identify the shape of the first objects in the row. Describe it. Circle the other shape(s) in the row that stack. **3.** Identify the shape of the first object in the row. Describe it. Circle the other shape(s) in the row that stack.

④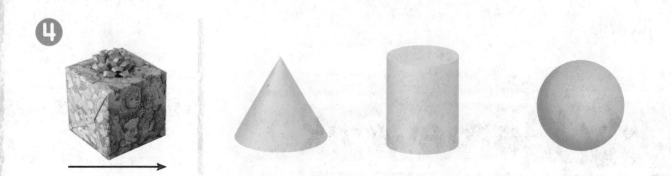

Vocabulary Check

⑤ **roll**

⑥ **stack**

⑦ **slide**

 Directions: 4. Identify the shape of the first object in the row. Describe it. Circle the other shape(s) in the row that slide. **5.** Identify the shapes. Circle the shapes that roll. **6.** Identify the shapes. Circle the shapes that can have another shape stacked on it. **7.** Identify the shapes. Circle the shapes that slide.

Math at Home Have your child name the solid shapes on the My Homework pages that roll, stack, and slide. Have your child imitate these movements.

Name _____

Check My Progress

Vocabulary Check

1 sphere

2 cylinder

Concept Check

3

 Directions: I. Draw an X on the objects that are not shaped like a sphere. **2.** Draw an X on the objects that are not shaped like a cylinder. **3.** Name the first shape in the row. Compare it to the shapes of the objects in the row. Circle the matching shape.

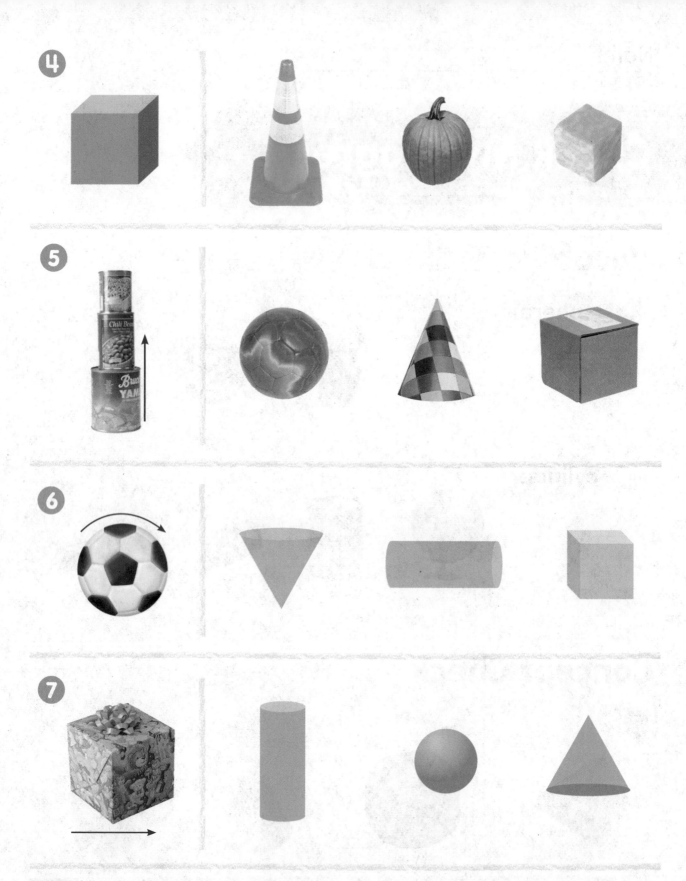

Directions: 4. Name the first shape in the row. Compare it to the shapes of the objects in the row. Circle the matching shape. **5.** Identify the first objects in the row. Describe them. Circle the other shape(s) in the row that stack. **6.** Identify the first object in the row. Describe it. Circle the other shape(s) in the row that stack. **7.** Identify the first object in the row. Describe it. Circle the other shape(s) in the row that slide.

712   Chapter 12

Name

Lesson 4
Problem Solving
STRATEGY: Act It Out

ESSENTIAL QUESTION
How do I identify and compare three-dimensional shapes?

What will stack on the cube?

Act It Out

 Teacher Directions: Name and describe the shapes in the castle. Trace the circles to show the shapes that you could stack on the cube. Use shapes to check the answer.

What will stack on the cylinder?

Act It Out

 Directions: Name and describe the shapes in the tower. Circle the shape(s) that you could stack on the cylinder. Draw an X on the shape(s) that could not stack on the cylinder. Use shapes to check the answer.

What will stack on the cone?

Act It Out

 Directions: Name and describe the shapes in the tower. Circle the shape(s) that you could stack on the cone. Draw an X on the shape(s) that could not stack on the cone. Use shapes to check your answer. Explain your answer to a friend.

What will stack on the cube?

Act It Out

 Directions: Name and describe the shapes in the tower. Circle the shape(s) that you could use to stack on the cube. Draw an X on the shape(s) that could not stack on the cube. Use shapes to check your answer.

Name _____

My Homework

What will stack on the cylinder?

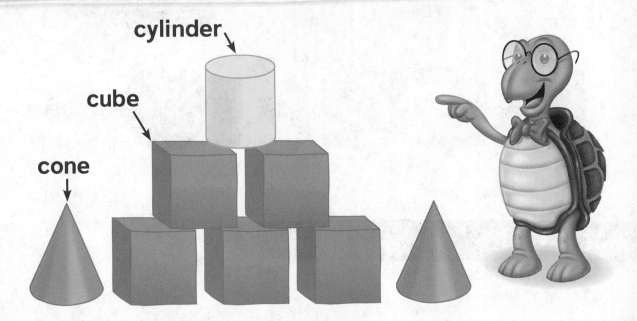

cylinder

cube

cone

Act It Out

Directions: Name and describe the shapes in the tower. Trace the circles to show the shapes that you could stack on the cylinder. Use a canned good, cube-shaped tissue box, and ball to check the answer.

What will stack on the cube?

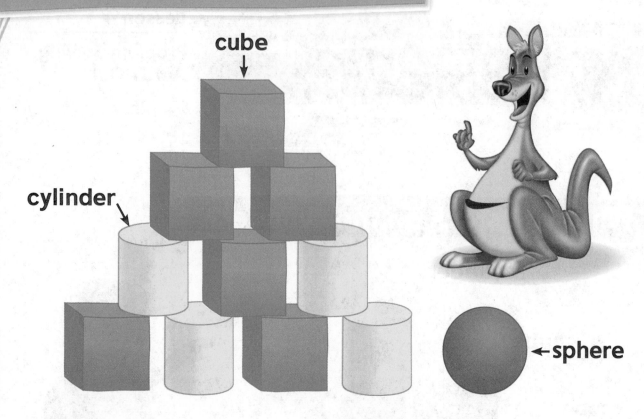

cube

cylinder

←sphere

Act It Out

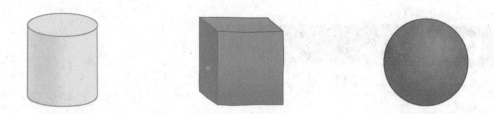

Directions: Name and describe the shapes in the tower. Circle the shapes that you could stack on the cube. Use a canned good, cube-shaped tissue box, and ball to check the answer.

Math at Home Take advantage of problem-solving opportunities during daily routines such as playing with blocks, cleaning up toys, or putting away groceries to identify objects that can stack, roll, or slide.

Name _____

Lesson 5
Model Solid Shapes in Our World

ESSENTIAL QUESTION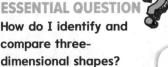
How do I identify and compare three-dimensional shapes?

Math in My World

Teacher Directions: Circle all the cones, cylinders, cubes, and spheres in the picture. Tell a classmate the name of each shape you found. Describe the shapes. Find objects in the classroom that are the same shapes.

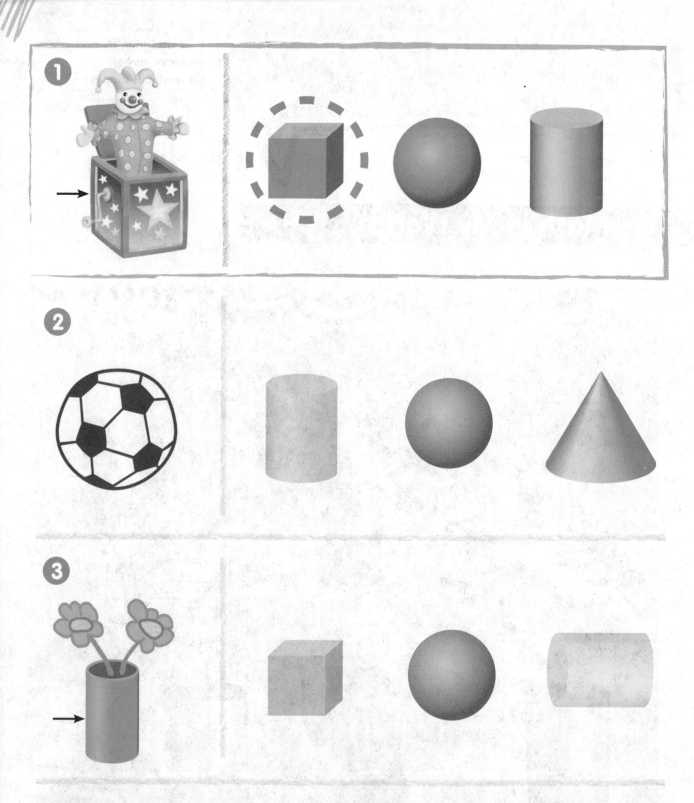

1

2

3

Directions: 1–3. Name the shape of the first object in the row. Describe it. Compare it to each shape in the row. Circle the matching shape.

Name

Independent Practice

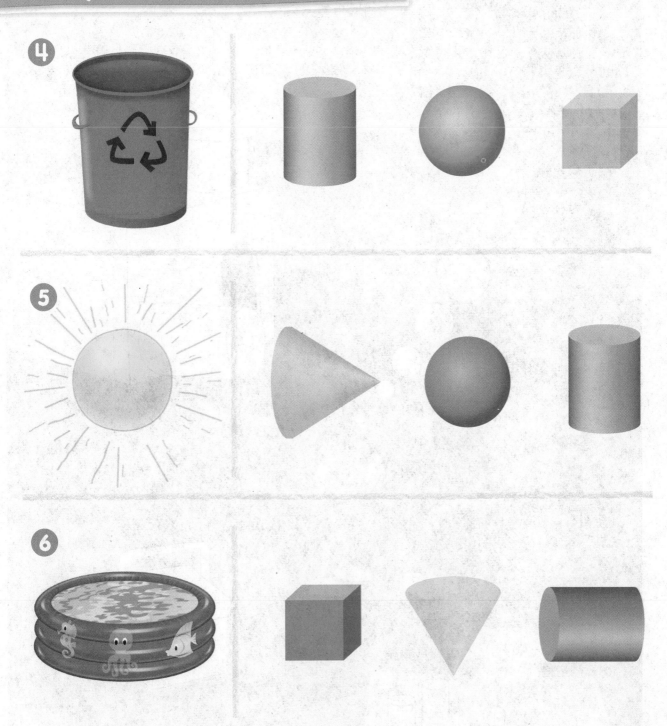

Directions: 4–6. Name the shape of the first object in the row. Describe it. Compare it to each shape in the row. Circle the matching shape.

Directions: 7. Color all the cubes yellow. Color all the cones red. Color all the spheres blue. Color all the cylinders orange. Describe the shapes.

Name _____

My Homework

Homework Helper

 eHelp

Need help? connectED.mcgraw-hill.com

1

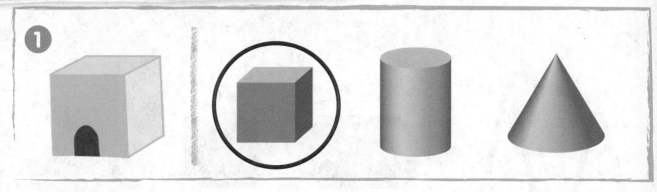

2

3

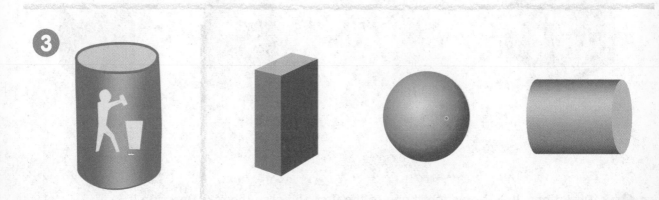

 Directions: 1–3. Name the shape of the first object in the row. Describe it. Compare it to each shape in the row. Circle the matching shape.

4

5

6

 Directions: 4–6. Name the shape of the first object in the row. Describe it. Compare it to each shape in the row. Circle the matching shape.

Math at Home Take a walk outside. Look for three-dimensional shapes. Draw pictures of the objects that are the shapes of cones, spheres, cylinders, and cubes.

Name

My Review

Vocabulary Check

sphere   **cube**   **cylinder**   **cone**

Directions: Draw a line from each vocabulary word to a shape that matches the word. Draw a box around the objects shaped like a sphere. Draw an X on the object shaped like a cube. Draw a line under the objects shaped like a cylinder. Circle the objects shaped like a cone.

Concept Check

 Directions: 1–2. Name the first shape in the row. Compare it to the shapes of the objects in the row. Circle the matching shape. **3.** Name the shape of the first objects in the row. Circle the other shape(s) in the row that stack. **4.** Name the shape of the first object in the row. Circle the other shape(s) in the row that stack.

Brain Builders

 Directions: 5. Name, identify, and describe the shapes. Color the spheres yellow. Color the cubes orange. Color the cones blue. Color the cylinders red.

Reflect

 Directions: Name each shape. Describe the shapes. Circle one of the shapes. Tell if the shape is solid or flat. Draw an object that is shaped the same as that shape. Describe the object to a classmate. Ask the classmate to name the shape of the object.

Performance Task

Brain Builders

Shapes and More Shapes

You can use solid three-dimensional shapes to identify the shape of real-world objects.

Part A

Match each real-world object to the three-dimensional solid shape.

   •   •

   •   •

   •   •

   •   •

Part B

Underline the shapes that will roll. Count them.
Write the number.

_____
- - - - - - -
_____

Part C

Circle the shapes that will stack. Count them.
Write the number.

_____
- - - - - - -
_____

Part D

Underline the shapes that will slide. Count them.
Write the number.

_____
- - - - - - -
_____

Glossary/Glosario

English   **Spanish/Español**

above

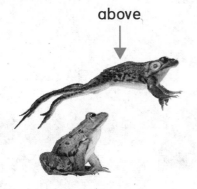

above

encima

encima

add

 = 5

3 ducks   2 more join   5 ducks in all

sumar

= 5

3 patos se unen 2 más 5 patos en total

afternoon

tarde

Aa

alike (same)

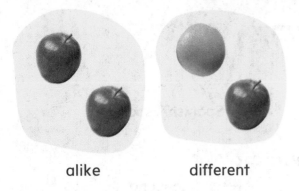

alike   different

igual

iguales   diferentes

are left

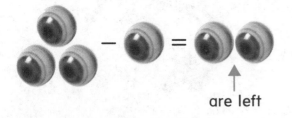

are left

quedan

quedan

Bb

behind

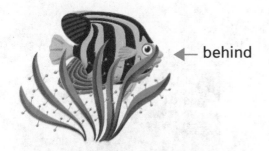

← behind

detrás

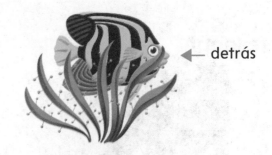

← detrás

below

↑
below

debajo

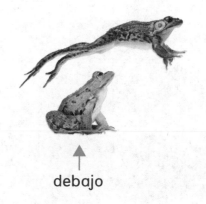

↑
debajo

beside

↑
The cat is beside the dog.

al lado

↑
El gato está al lado del perro.

calendar

			April			
Sunday	Monday	Tuesday	Wednesday	Thursday	Friday	Saturday
		1	2	3	4	5
6	7	8	9	10	11	12
13	14	15	16	17	18	19
20	21	22	23	24	25	26
27	28	29	30			

calendario

			abril			
domingo	lunes	martes	miércoles	jueves	viernes	sábado
		1	2	3	4	5
6	7	8	9	10	11	12
13	14	15	16	17	18	19
20	21	22	23	24	25	26
27	28	29	30			

Cc

capacity

holds more   holds less

capacidad

contiene más   contiene menos

circle

círculo

compare

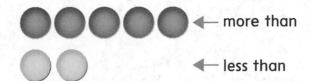

← more than

← less than

comparar

← más que

← menos que

cone

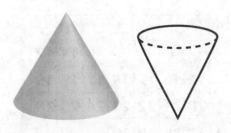

cono

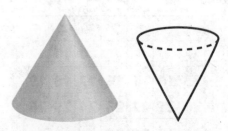

corner

corner (vertex)

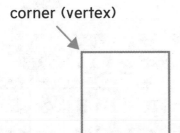

esquina

esquina (vértice)

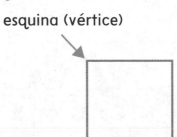

count

1	2	3
one	two	three

contar

1	2	3
uno	dos	tres

cube

 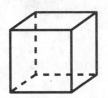

cubo

 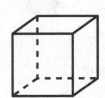

cylinder

cilindro

Dd

day

día

different

different   alike

diferente

diferentes   iguales

Ee

eight

ocho

eighteen

dieciocho

eleven

once

equals sign (=)

$$4 + 1 = 5$$

↑
equals

signo igual (=)

$$4 + 1 = 5$$

↑
es igual a

equal to

igual a

Ee

evening

noche

fifteen

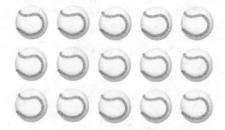

quince

five

cinco

four

cuatro

fourteen

catorce

Gg

greater than

mayor que

Hh

heavy (heavier)

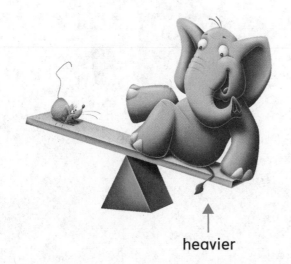

heavier

pesado (más pesado)

más pesado

height

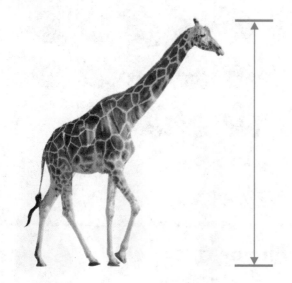

altura

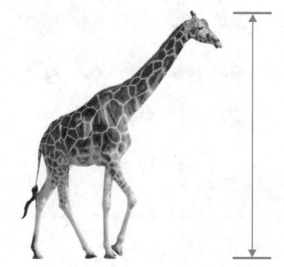

hexagon

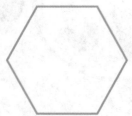

hexágono

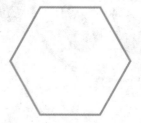

holds less

holds less

contiene menos

contiene menos

holds more

holds more

contiene más

contiene más

holds the same

holds the same

contiene la misma cantidad

contiene la misma cantidad

Ii

in all

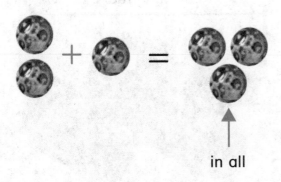

in all

en total

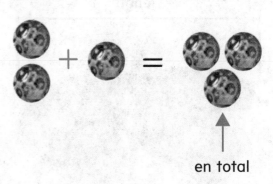

en total

in front of

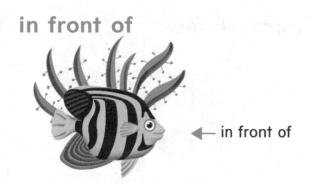

← in front of

en frente de

← en frente de

 Jj

join

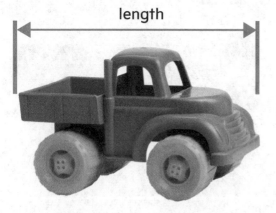

3 birds and 2 birds join.

juntar

Hay 3 aves y se les juntan 2 más.

 Ll

length

length

longitud

longitud

less than

menor que

light (lighter)

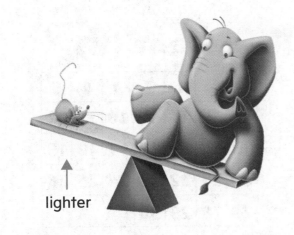

lighter

liviano (más liviano)

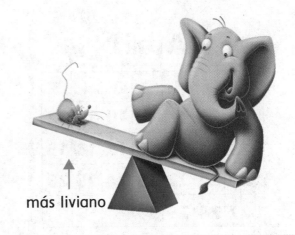

más liviano

long (longer)

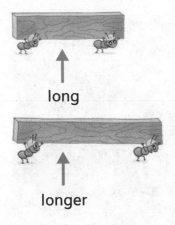

long

longer

largo (más largo)

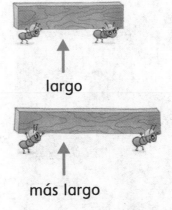

largo

más largo

minus sign (−)

$$5 - 2 = 3$$

minus

signo menos (−)

$$5 - 2 = 3$$

menos

month

month

mes

mes

morning

mañana

next to

The cat is next to the dog.

junto a (al lado)

El gato está junto al perro.

nine

nueve

nineteen

diecinueve

number

3

tells how many

número

3

dice cuántos hay

Online Content at connectED.mcgraw-hill.com

Glossary/Glosario

one

uno

ordinal numbers

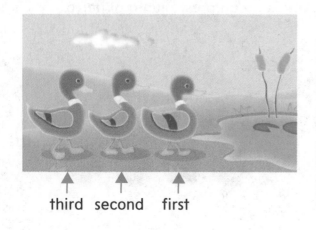

third   second   first

números ordinales

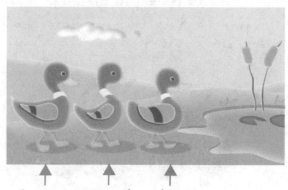

tercero   segundo   primero

pattern

A, B, A, B, A, B

repeating pattern

patrón

A, B, A, B, A, B

patrón que se repite

plus sign (+)

$$5 + 2 = 7$$

↑
plus

signo más (+)

$$5 + 2 = 7$$

↑
más

position

above

below

posición

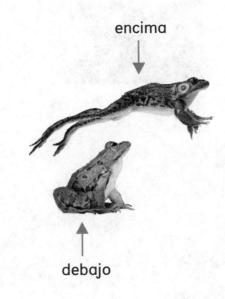

encima

debajo

Rr

rectangle

rectángulo

Rr

repeating pattern

repeating pattern

patrón que se repite

patrón que se repite

roll

rodar

round

round · · · not round

redondo

redondo · · · no redondo

same number

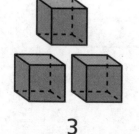
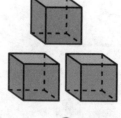

3 · · · 3

same number

el mismo número

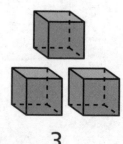

3 · · · 3

el mismo número

separate

separar

seven

siete

seventeen

diecisiete

shape

figura

Ss

short (shorter)

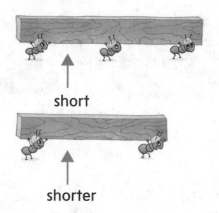

short

shorter

corto (más corto)

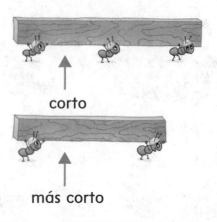

corto

más corto

side

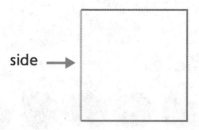

side →

lado

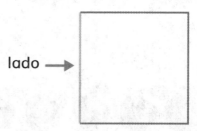

lado →

six

seis

sixteen

dieciséis

size

small   medium   large

tamaño

pequeño   mediano   grande

slide

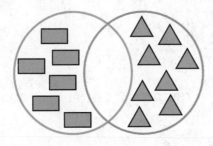

deslizar

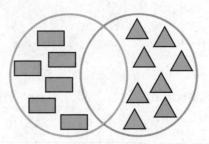

sort

sorted or grouped by shape

ordenar

ordenados o agrupados por su forma

sphere

esfera

square

cuadrado

stack

pila

straight

straight   not straight

recto

recto   no recto

subtract (subtraction)

5 take away 3 is 2. 2 are left.

restar (resta)

Si a 5 le quitamos 3, quedan 2.

tall (taller)

taller

alto (más alto)

más alto

ten

diez

thirteen

trece

three

tres

three-dimensional shape

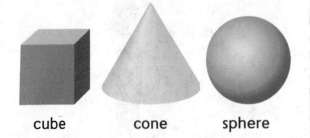

cube   cone   sphere

figura tridimensional

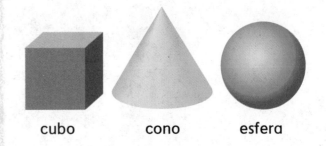

cubo   cono   esfera

today

hoy

tomorrow

mañana

triangle

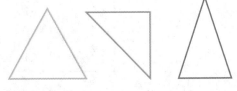

triángulo

twelve

doce

twenty

veinte

two

dos

Tt

two-dimensional shape

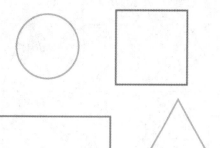

figura bidimensional

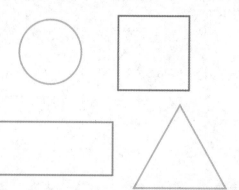

vertex

vertex
(corner)

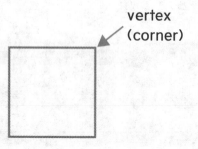

vértice

vértice
(esquina)

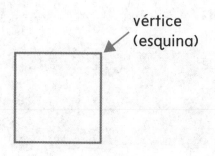

week

week

semana

semana

weight

heavy   light

peso

pesado   liviano

year

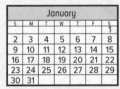

January

S	M	T	W	T	F	S
						1
2	3	4	5	6	7	8
9	10	11	12	13	14	15
16	17	18	19	20	21	22
23	24	25	26	27	28	29
30	31					

February

S	M	T	W	T	F	S
		1	2	3	4	5
6	7	8	9	10	11	12
13	14	15	16	17	18	19
20	21	22	23	24	25	26
27	28					

March

S	M	T	W	T	F	S
		1	2	3	4	5
6	7	8	9	10	11	12
13	14	15	16	17	18	19
20	21	22	23	24	25	26
27	28	29	30	31		

April

S	M	T	W	T	F	S
					1	2
3	4	5	6	7	8	9
10	11	12	13	14	15	16
17	18	19	20	21	22	23
24	25	26	27	28	29	30

May

S	M	T	W	T	F	S
1	2	3	4	5	6	7
8	9	10	11	12	13	14
15	16	17	18	19	20	21
22	23	24	25	26	27	28
29	30	31				

June

S	M	T	W	T	F	S
			1	2	3	4
5	6	7	8	9	10	11
12	13	14	15	16	17	18
19	20	21	22	23	24	25
26	27	28	29	30		

July

S	M	T	W	T	F	S
					1	2
3	4	5	6	7	8	9
10	11	12	13	14	15	16
17	18	19	20	21	22	23
24	25	26	27	28	29	30
31						

August

S	M	T	W	T	F	S
	1	2	3	4	5	6
7	8	9	10	11	12	13
14	15	16	17	18	19	20
21	22	23	24	25	26	27
28	29	30	31			

September

S	M	T	W	T	F	S
				1	2	3
4	5	6	7	8	9	10
11	12	13	14	15	16	17
18	19	20	21	22	23	24
25	26	27	28	29	30	

October

S	M	T	W	T	F	S
						1
2	3	4	5	6	7	8
9	10	11	12	13	14	15
16	17	18	19	20	21	22
23	24	25	26	27	28	29
30	31					

November

S	M	T	W	T	F	S
		1	2	3	4	5
6	7	8	9	10	11	12
13	14	15	16	17	18	19
20	21	22	23	24	25	26
27	28	29	30			

December

S	M	T	W	T	F	S
				1	2	3
4	5	6	7	8	9	10
11	12	13	14	15	16	17
18	19	20	21	22	23	24
25	26	27	28	29	30	31

año

enero

d	l	m	m	j	v	s
						1
2	3	4	5	6	7	8
9	10	11	12	13	14	15
16	17	18	19	20	21	22
23	24	25	26	27	28	29
30	31					

febrero

d	l	m	m	j	v	s
		1	2	3	4	5
6	7	8	9	10	11	12
13	14	15	16	17	18	19
20	21	22	23	24	25	26
27	28					

marzo

d	l	m	m	j	v	s
		1	2	3	4	5
6	7	8	9	10	11	12
13	14	15	16	17	18	19
20	21	22	23	24	25	26
27	28	29	30	31		

abril

d	l	m	m	j	v	s
					1	2
3	4	5	6	7	8	9
10	11	12	13	14	15	16
17	18	19	20	21	22	23
24	25	26	27	28	29	30

mayo

d	l	m	m	j	v	s
1	2	3	4	5	6	7
8	9	10	11	12	13	14
15	16	17	18	19	20	21
22	23	24	25	26	27	28
29	30	31				

junio

d	l	m	m	j	v	s
			1	2	3	4
5	6	7	8	9	10	11
12	13	14	15	16	17	18
19	20	21	22	23	24	25
26	27	28	29	30		

julio

d	l	m	m	j	v	s
					1	2
3	4	5	6	7	8	9
10	11	12	13	14	15	16
17	18	19	20	21	22	23
24	25	26	27	28	29	30
31						

agosto

d	l	m	m	j	v	s
	1	2	3	4	5	6
7	8	9	10	11	12	13
14	15	16	17	18	19	20
21	22	23	24	25	26	27
28	29	30	31			

septiembre

d	l	m	m	j	v	s
				1	2	3
4	5	6	7	8	9	10
11	12	13	14	15	16	17
18	19	20	21	22	23	24
25	26	27	28	29	30	

octubre

d	l	m	m	j	v	s
						1
2	3	4	5	6	7	8
9	10	11	12	13	14	15
16	17	18	19	20	21	22
23	24	25	26	27	28	29
30	31					

noviembre

d	l	m	m	j	v	s
		1	2	3	4	5
6	7	8	9	10	11	12
13	14	15	16	17	18	19
20	21	22	23	24	25	26
27	28	29	30			

diciembre

d	l	m	m	j	v	s
				1	2	3
4	5	6	7	8	9	10
11	12	13	14	15	16	17
18	19	20	21	22	23	24
25	26	27	28	29	30	31

Online Content at connectED.mcgraw-hill.com

Glossary/Glosario   GL27

yesterday

yesterday   today

Sunday	Monday	Tuesday	Wednesday	Thursday	Friday	Saturday
		1	2	3	4	5
6	7	8	9	10	11	12
13	14	15	16	17	18	19
20	21	22	23	24	25	26
27	28	29	30			

April

ayer

ayer   hoy

domingo	lunes	martes	miércoles	jueves	viernes	sábado
		1	2	3	4	5
6	7	8	9	10	11	12
13	14	15	16	17	18	19
20	21	22	23	24	25	26
27	28	29	30			

abril

Zz

zero

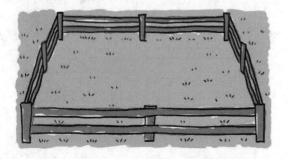

cero

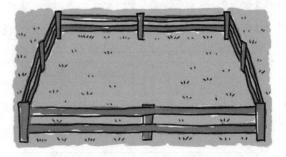